WINDPUMPS

WINDPUMPS

A guide for development workers

Peter Fraenkel, Roy Barlow,
Frances Crick, Anthony Derrick
and Varis Bokalders

Intermediate Technology Publications in association with the Stockholm
Environment Institute 1993

Published by Intermediate Technology Publications Ltd
103-105 Southampton Row, London WC1B 4HH, UK.

ISBN 1 85339 126 3

Printed by The Russell Press, Radford Mill, Norton Street, Nottingham NG7 3HN, UK.

Foreword

Windpumps represent a well-developed technology which has been applied world-wide for centuries. Windpumps have been used for drinking water supply, irrigation and drainage of wetlands. They have proved to be invaluable in the development of livestock industries in dry areas and are well-suited for meeting many types of agricultural water demand.

Historically, windpumps have been used to develop the great plains of the USA, the outback of Australia, the Patagonian Plains in Argentina, Jiantsu Province in China, south-eastern Russia, parts of South Africa and many other places. They have also been used extensively in many island societies like Mallorca, Crete, the Canary Is., Cape Verde Is., and the West Indies.

Facing competition from other energy sources, there was a decline in the number of windpumps after the Second World War. However, both fluctuating oil prices, and considerable development of windpump technology during the 1970s and 1980s, are now again changing the situation. Windpumps are again becoming a competitive renewable energy technology, considered to be environmentally benign, that can be useful for water pumping in non-electrified rural areas, mainly in developing countries.

Windpumps are well suited for water pumping in wind regimes with more than 2.5 to 3m/s mean windspeed. However, they are vulnerable to very strong winds and can be difficult to use in areas that frequently have storms or hurricanes. When an area is electrified, windpumps are usually rapidly replaced by electric pumps that are cheaper and easier to maintain. When diesel fuel is cheap and easily available, diesel pumps can be an attractive alternative if regular servicing and maintenance can be provided. However, there are still large areas of the world where there is no electricity and where windpumps can often be the best solution for water pumping.

The purpose of this book is to make it easier for anyone interested in acquiring a windpump to make the right choice and to take into account all the factors involved, both technical and contextual.

Lars Kristoferson
Stockholm Environment Institute
April 1993

Preface

This book is the result of a co-operative project involving I T Power Ltd, the Stockholm Environment Institute (SEI) and the Swedish Mission Council (SMC), and has been sponsored by the Swedish International Development Authority. Its origin stems from the needs of SMC field staff who have found much of the information currently available on windpumping to be fragmented and often incompatible.

The Stockholm Environment Institute, which has close working relations with SMC in the field of renewable energy, runs an information programme on renewable energy for development which has resulted in a series of publications and seminars.

I T Power has had substantial experience in the field of windpumping and designed the IT windpump now manufactured in several developing countries. I T Power works more broadly in all areas of renewable energy application.

This guidebook aims to provide information on the technology, economics, performance, experiences and potential of windpumps for water pumping in developing countries.

The objective of the guide is to assist development workers and project managers working in water supply, rural development, farming and livestock sectors to decide:

- If windpumps are suitable for their application

- Which type of windpump would be most suitable

- How to implement a project using windpumps including:

    ✓ assessing the wind resource
    ✓ assessing the water resource
    ✓ evaluating the relative economics
    ✓ sizing and specifying a windpump
    ✓ procurement and installation
    ✓ maintenance and operation

This is the third in the series of guides for development workers on energy technologies for sustainable development. Other guides available are:

Solar Photovoltaic Products
First ed. January 1989 (by A. Derrick *et al*)
Updated January 1991
ISBN 1 85339 091 7

Micro-hydro Power
January 1992 (by P L Fraenkel *et al*)
ISBN 1 85339 029 1

Contact persons are:

Peter Fraenkel	Varis Bokalders	Karl-Erik Lundgren
I T Power Ltd	Stockholm Environment Institute	Swedish Mission Council
The Warren	Box 2142	Office for international
Bramshill Road	Jarntorget 84	Development Co-operation
Eversley	103 14 Stockholm	Tegnergatan 34 n.b.
Hampshire, RG27 0PR	Sweden	113 86 Stockholm
UK		Sweden
Fax: +44 734 730820	Fax +46 8 723 0348	Fax +46 8 31 58 28

Contents

Acknowledgements

The authors wish to thank the Swedish International Development Authority, SIDA, for supporting the production of this guide and providing the finance for the research work.

The review comments from Rene Karottki and Lars Yde of the Danish Folkecenter are gratefully acknowledged, as are the comments and suggestions received from colleagues of the authors.

The guide could not have been produced without the kind assistance from the international windpump industry in providing product data and photographs. Thanks are also due to T. Lindsay Baker of the Windmillers' Gazette for the historical photographs in Figures 2.8 and 2.9, and to Kirk Stokes of Neos Corporation for Figure 3.10. Figures 4.11 to 4.13 are reproduced from *Wind Energy* by T Kovarik by permission of Prism Press.

Thanks also go to Nigel Wood of the Royal Meteorological Office for advice on wind-flow over topography, and to Kathleen Glancey for her help in laying out the text and buyers' guide.

NOTICE

Introduction

1

1.1 Why Windpumps?

Windpumps are wind-driven machines for pumping water.

Windpumps are considered to be an environmentally benign technology for reliable pumping of water in rural areas. They can be cost competitive (compared to diesel or solar pumps for example) in many locations.

Windpumps can be manufactured in small engineering workshops in most developing countries thereby developing indigenous skills, increasing employment, creating economic activity and offsetting imports (of diesel fuel and diesel pumps). In some cases (for example in Argentina and Kenya) windpump manufacture is also an export industry.

The technical committee on wind energy at the United Nations Conference on New and Renewable Sources of Energy in 1981 foresaw 'wind energy playing a significant role in pumping water for households, animal husbandry, irrigation and drainage in rural areas of developing countries'.

The committee concluded that that the potential for windpumps in the developing world would run into 'many millions'.

In the ten years since then, however, the dissemination of windpumps in the developing world has been fairly limited. One reason cited for this is lack of appropriate information on the technology and experiences for consideration by potential users and procurement agencies. This handbook aims to overcome this problem.

Figure 1.1 Tens of thousands of this *Fiasa* windpump have been manufactured in Argentina, many for export

1

The principal benefits of windpumps are:

- Windpumps are the most economic method of pumping water in rural areas where the average windspeed is greater than about 3m/s, and no grid power is available

- Windpumps have no fuel requirements: unlike engine-driven pumps

- Windpumps are a clean technology as there is no harmful pollution as the result of their use

- Windpumps are among the most robust of pumping systems given regular maintenance

- Windpumps have a long life potential, typically 20 years for a well-made and well-maintained machine

- Windpumps can be locally manufactured in most developing countries

Figure 1.2   Rotor ring assembly during installation (*Tawana*, Pakistan)

1.2   Non-Technical Aspects of Water Supply Projects

Water provision is one of the most critical aspects of development, but there are many factors that make its realization far from straightforward.

It is not only technological barriers that must be overcome, but there are many non-technical considerations that must be taken into account.

The manner in which water is supplied and the way a project is planned, implemented and managed are important subjects.

'Appropriateness' is less a technological issue than an organizational issue, but until recently technology has tended to dominate thinking in the area of water development.

In summary, the following socio-economic considerations are valid for almost all water development projects:

- The objectives of a water supply project must be clearly understood

- If the water supply is for a community, that community must feel some form of ownership and participate fully in the planning and implementation of the project

- Donors should not merely aim to finance initial costs, but should also consider the recurrent costs that the recipient community may have to bear

- Politics and relations between donors, governments, institutions, groups and individuals must be clearly understood

- If the water supply is for private individuals, questions of access, maintenance, operation and costing are quite simple, and management is much more straightforward

- If the water supply is for institutions, then the questions of maintenance control and management must be driven by institutional presence, and commitment to that presence

The Evolution of Windpumps

2

2.1 The History of Windpower

It is known that the ancient Egyptians used windpower as long as 5000 years ago to propel boats. The actual date when windpower was first used on land to power machinery remains obscure, although it is recorded that Hero of Alexandria built a wind-powered organ in 260 BC. However, it was only a curiosity, and it is believed that the ancient Greeks never used wind energy for any practical use. Wind-powered prayer-wheels, used by Buddhists in central Asia, were first described by Chinese travellers about 400 AD.

Some historians claim that windmills existed long before this, although their theories are not universally accepted. Stanley Freese (*Windmills and Mill-wrighting*, 1971) claims that windmills, for lifting water, are mentioned in the Indian book, *Arthastra of Kanti-lya* (400 BC). A. Flettner says in his book (*The Story of the Rotor*, 1926) that the Babylonian Caesar Hammurabi planned wind-powered irrigation in 1600 BC. David Rittenhouse Ingels (*Wind Energy*, 1978) claims that there is a picture of a windmill on a Cinese vase from 3000 BC.

The first known reference to a windmill occurs in the Arab historian Tabari's anecdote (written in the 850s) about the Persian crafts-men-slave Abu Lulua, from Sistan (Seistan province on the border of Iran and Afghanistan), who in 644 AD boasted to Arab Caliph Omar that he could construct a mill turned by the wind. He murdered the Caliph the day after this recorded conversation.

The work of Al-Masudi, completed in 943 AD, recounts the tale of Abu-Lulua, but also described Sistan as 'famous for the industry with which the wind is used to turn mill-stones and to draw irrigation water from

wells'. It is assumed that these first Sistani windmills powered so-called Persian Wheels (a chain of pots used to raise water from wells) common in Middle Asia and using gearing well known from the Vitruvian water wheels.

The most important medieval reference to Persian windmills, notable for its wealth of technical detail, is to be found in the writings of al-Dimashgi c.1300 (see Figure 2.1): 'When building mills that rotate by the wind, they (in Sistan) erect a high building like a minaret, or they take a high mountain or hill, or tower of

Figure 2.1 Seistanian windmill, c.1300, from the writings of al-Dimashgi

a castle. They build one building on top of another. The upper structure contains the mill that turns and grinds; the lower one contains a wheel rotated by the enclosed wind. After they have completed the structures, as shown on the drawing, they make four slits in the walls, suitable for the entry of the wind, which penetrates the mill house from whichever direction the wind may blow'. Part of its interest lies in the fact that the stones were placed above the wind-wheel, and that the surrounding walls had slits on all sides so that the wind could operate the mill from all directions.

There are still horizontal windmills in Seistan and Khorasan provinces at the border between Iran and Afghanistan. There are about 50 remaining, some of which may still be in use today. They utilize the steady winds that blow for 120 days from NNE in Khorasan and from NNW in Seistan. As these winds blow from mid-June until mid-October in the same direction, the windmills are fixed and can not be turned to utilize winds from other directions. Vertical sails attached to the spokes of a tall drum cause it to rotate horizontally, so that its central axle drives the runner-stone in the mill-house below (see Figure 2.2). However, these mills have the stones below and the wind-wheel above.

The same kind of windmills that are common in Iran and Afghanistan have also been recorded in Russia (Samarkand and Omsk) and Pakistan (Baluchistan and Damaun).

The twelfth century Sicilian Muslim geographer al-Idrisi recorded that 'the inhabitants of Malaya have mills turned by the wind where they grind into flour rice, wheat and other cereals'. Two hundred years later Ibn al-Wardi, a Syrian encyclopedist, referred to the Indonesian Island Halathi: 'there are also windmill (placed upon rafts and floating) on the water'. Transmission of technology at that early date was probably due to Persian sailors. It is assumed that the windmills were vertical-axis and of Persian origin.

The Chinese had embassies to Samarkand in 1219 and to Herat in 1414 and reported back

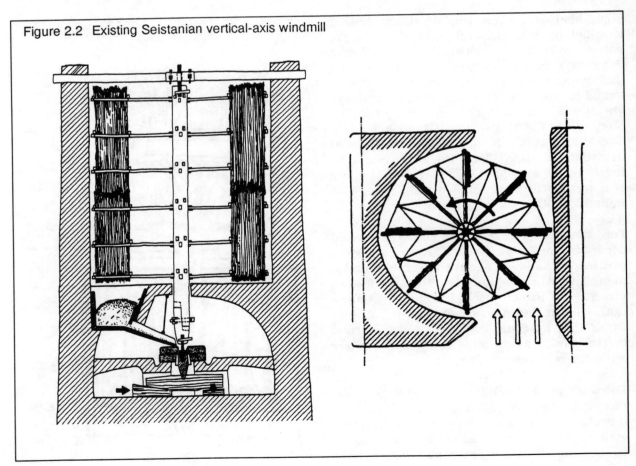

Figure 2.2 Existing Seistanian vertical-axis windmill

to China the existence of windmills in Persia. It is belie-ved that the Chinese wind-pumps, which were vertical axis and only recorded after the 12th century were in-spired by the windmills in Sistan. These kind of vertical-axis windpumps still operate salt-works in China today (see Figure 2.3).

Vertical-axis windmills were also used in Crimea among the Armenians (especially in Eupatoria and Pallas) and it is interesting to note that the larger part of old Armenia once belonged to Persia. By the middle of the nineteenth century only one vertical-axis mill remained in Eupatoria, but about 200 horizontal-axis windmills were found there.

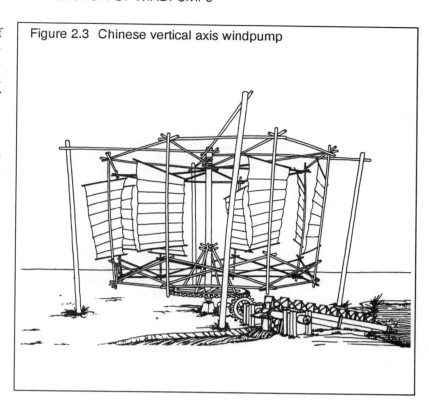

Figure 2.3 Chinese vertical axis windpump

From Asia to Europe

The Muslim expansion during the reign of the Umayyad Caliphs that ruled from Damascus in Syria, led to the conquest of Spain in the west in 711 and of Iran in the east. By 750 there was one rule from the Iberian peninsula to the border of China, that was kept together by Islam as a religion and used Arabic as its administrative language. It is therefore quite probable that windmill technology was trans-ferred to Europe during this period. The cultural role of Islamic Spain should not be underestimated. The Umayyades ruled Spain until 1031 and the last Muslim ruler was thrown out of Spain in 1492.

In Spain windmills had already been men-tioned in the works 'The Golden Meadows' by Al'-Mas Udï who lived 912-957 (according to Julio Caro Baroja). The Arabic poet Ibne Mocana from Alcabidache in Portugal was already writing poems about windmills in the 11th century. These are indications that the windmill came to Europe with the Muslims via the Iberian peninsula in the middle ages. After the Muslims were driven out of Northern Spain there are no known records of wind-mills in Spain until 1330, or in Portugal until 1303.

On the Greek Island of Karpathos there are ruins of primitive windmills of the same type that al-Dimashgi described in 1300. These were fixed vertical-axis windmills where the sails are below the stones. On the same island there are also fixed horizontal-axis windmills so that it is possible to follow the development from vertical- to horizontal-axis designs.

Windmill expansion in Europe

During the time of the crusades (first crusade 1096-1099, second crusade 1147-1149 and third crusade 1189-1191) the knowledge of windmills had spread across Europe. Docu-ments prove that windmills existed in France (Normandy) in 1180, Belgium (Opzullik) 1193, England (Shineshead, Lincolnshire) 1181 and in Syria (ruled by the crusaders) 1190. Howe-ver, pilgrims travelling to Santiago de Com-postela in Spain might well have introduced windpower technology at an earlier date. Unproven tradition talks of windmills in Flan-ders in 1001, France 1090 and England 1066.

Later the expansion of windmills followed in the track of the establishment of the monas-teries (Cistercians) and colonization. Windmills are first recorded in Holland in 1240 (Merum), Germany in 1222 (Cologne) and 1234 (Schlesvig Hollstein), Italy in 1237 (Siena), Greece in 1239, Denmark in 1259, Sweden in 1330 (Skåne), Latvia in 1330 (Riga), Estonia in 1353, Finland in 1463, Hungary in 1550, Romania (Dobrogea) in 1585, Brazil in 1576 (Rio de Janeiro), USA in 1621 (Virginia), Russia in 1622 (Archangelsk), Barbados in 1651 and South Africa (Cape Town) by the end of the 17th century.

Technical development of the windmill

Various forms of horizontal-axis windmill developed in Europe (see Figure 2.5). The first illustration of an European windmill is in the English Windmill Psalter 1271, and shows a vertical post mill (where the whole mill turns on a post). The post mill was quite quickly followed by the tower mill (only the top section with the sails turns), first illustrated on a tapestry from 1390 in Nürnberg in Germany. The smock mill or 'Dutch windmill' was invented by Leif Andveitz in Holland (1573). In this design the cap with the sails is turned, but the gear wheels and mechanisms are fixed in the mill body.

By the 18th century technical improvements began to appear. In 1745 Edmond Lee patented the self-regulating wind machine. This had a fantail, a miniature windmill, enabling the main windmill to turn in the wind automatically.

In 1752-3 John Smeaton conducted scientific experiments with windmill sails (see Figure 2.4), and introduced cast-iron parts. In 1772 Andrew Meikle made a plan for a shuttered sail. Thomas Mead patented a regulator in 1787, and in 1789 Stephen Hooper put the ideas together in the *Patent Sail*. By 1800 the windmill had reached a technological peak. Much of the engineering know-how developed while perfecting windmills led to the rapid development of the steam engine by Watt and others in the late 18th century and hence the industrial revolution.

By the end of the 19th century there were reportedly some 10,000 windmills in the UK, 18,000 in Germany and several thousand in many other countries thoughout Europe.

Windpumps

Windpumps *Wipmolen* were first used in Europe in Holland (1408 in Alkmaar), probably using a scoop-wheel for pumping. By 1600 the wind-driven scoop-wheel was to be seen in the low-lying lands of western Europe and Britain (see Figure 2.5). The wind-powered Archimedes screw was introduced in Holland in 1634 (see Figure 2.6).

The first use of the annular sail, in which all the shutters are assembled radially, was made by Henry Chopping and Richard Rufflein in England 1838, and this concept was exploited for windpumps. The advantage of the shuttered sail lay in its ability to work

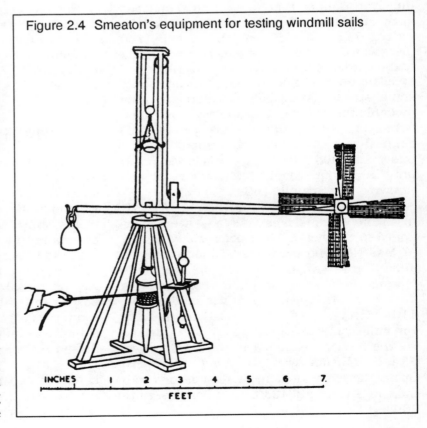

Figure 2.4 Smeaton's equipment for testing windmill sails

INCHES 1 2 3 4 5 6 7.

FEET

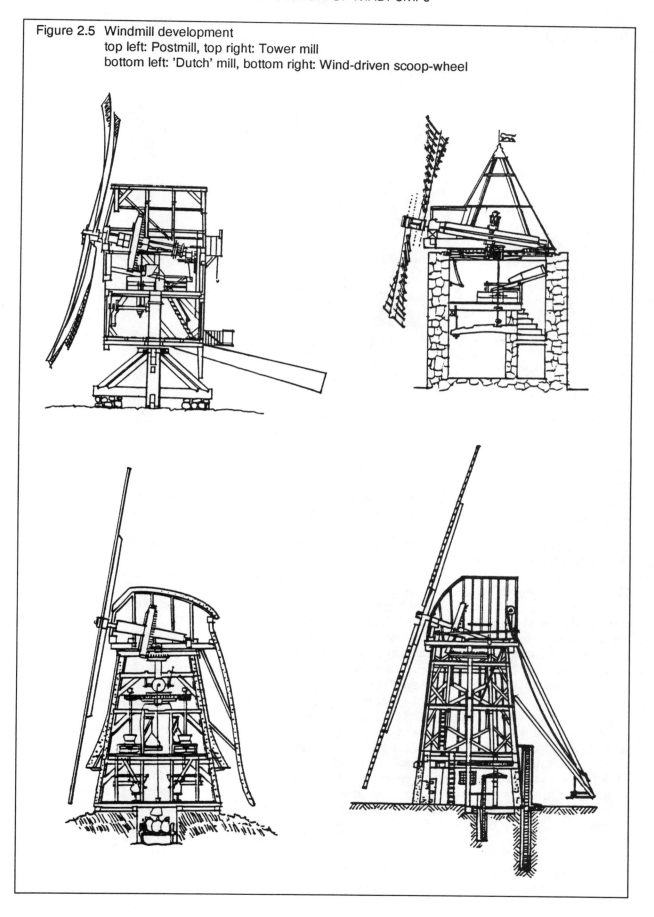

Figure 2.5   Windmill development
top left: Postmill, top right: Tower mill
bottom left: 'Dutch' mill, bottom right: Wind-driven scoop-wheel

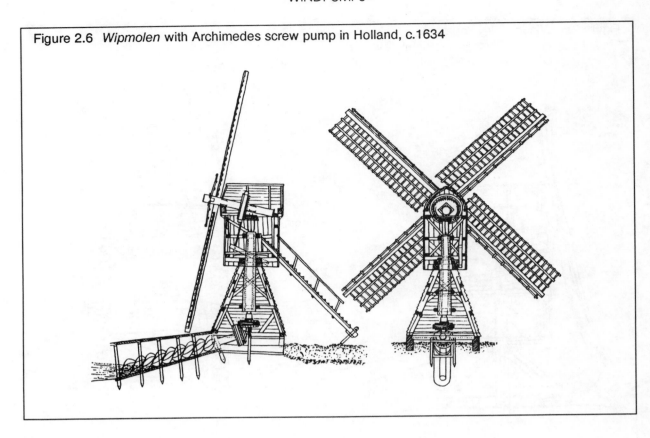

Figure 2.6 *Wipmolen* with Archimedes screw pump in Holland, c.1634

automatically, both in its control under variations in windspeed and its ability to face into the eye of the wind when the direction changed.

Windpumps in the USA

The development of the windpump shifted from Europe to the United States during the nineteenth century. As settlers moved across the Great Plains of the USA during the 1850s moving large herds of cattle with them, one of their biggest problems was the provision of water in adequate quantities to sustain their livestock in areas having plentiful grazing but limited surface water. Cattle can only graze where they are within reach of the water they need to survive. Pumping water by hand from numerous small wells and creeks was too onerous a task for the limited numbers of frontiersmen, so they started to improvise windmills to pump water for the cattle, utilising the steady breezes that blow across the open prairies.

All manner of different types of windpump were built in the USA from the 1850s

onwards. Some were quite bizarre (ref: *The Homemade Windmills of Nebraska*, 1898). Some were horizontal-axis machines but many were fixed in orientation like the original Asian windmills, in areas where the winds tended to blow from the same direction.

A popular type, because they were easy to build, was the *Jumbo Mill* in which a large paddle, like the paddle from a contemporary paddle steamer of that time, was mounted with its shaft horizontal with a wooden box around its lower half. The wind blowing across the top would turn it much in the same was as an undershot waterwheel works. The efficiency was very poor but such machines were cheap and simple to build. Sometimes Jumbo mills were even built into the roofs of barns.

One popular type was the so-called 'battle-ax' windpump, named because of its axe-shaped wooden sails driving the horizontal shaft. Like the Jumbo this depended on prevailing winds as it was fixed in orientation. A fairly similar design of simple windpump is still used today for brine pumping in salt workings in the Cape Verde Islands.

Figure 2.7 Cretan Sail windpump in Ethiopia

Soon the cleverer designs, which generally involved a windmill head capable of orienting to face the wind from any direction, went into production in the US by fledgeling agricultural engineering companies, as construction materials were hard to come by on the open grasslands. By the 1880s there was a major windpump manufacturing industry in the USA, involving numerous companies. Figure 2.8 shows several historic designs which have been restored and erected near Stinnett in Texas, USA.

Apart from the cowboys with their cattle, the railways were spreading across America and regular water tanks were needed to replenish the inefficient and thirsty steam locomotives. It soon became a common sight to see a windpump beside the tank-stand and water-crane at rural railway sidings in the mid-West of the USA. Similarly, the colonial powers started pushing railways across the arid regions of Africa, Latin America and Australia. They copied the American practice, commonly installing windpumps to keep water tanks for replenishment of locomotives filled.

Figure 2.8 A collection of historic windpumps gathered and restored by J B Buchanan, Stinnett, Texas, USA

By the end of the nineteenth century many settlements in the arid and semi-arid regions of the world had a small group of multi-bladed windpumps not only providing water for livestock but also for the human settlements.

In many ways the ubiquitous windpump, commonly seen in the background in movies of the Wild West or the Australian outback was perhaps the least thought about but most essential item of technology behind the exploitation of such regions. No windpumps would have meant no water and hence no livestock or livelihood for the inhabitants.

The first commercially successful self-governing 'American Farm Windpump' was invented in 1854 by a New England mechanic named Daniel Halladay. It had a rotor shape a bit like a flower that folded in on itself as the wind speed increased, thus presenting a lesser area to the wind. They served the farms in the mid-west and the railways in the US. Leonard Wheeler in 1866 invented a windpump (Eclipse) with two vanes, one that held the sails towards the wind and a smaller one that pushed the sails out of the wind as wind velocities increased. Very quickly the paddle-shaped blades were supplanted by wheels composed of thin wooden blades fixed to wooden rims.

The sails transmitted power to the pump rods at the head of the tower where the shaft turned a crank and thereby achieved the reciprocal motion needed for the pump. Bearings were mostly made of babbitt metal and worked well when frequently lubricated with oil.

In those days reliable lubricating greases had not been developed. The next step in the evolution of the windmill lubrication was the use of automatic lubricators; containers which fed oil gradually to the bearings to extend the periods for which the machine could operate reliably unattended.

In fact, much of the development of the windpump was aimed at making it increasingly reliable, robust and long-lasting and reducing the frequency of human intervention, bearing in mind that many livestock watering points are in remote and inaccessible areas.

The first all-steel windpumps

The early windpumps were made of wood. By the 1870s a new style of mill made from iron and steel began appearing. But two decades passed before substantial numbers of steel mills came into general production. The situation changed during the 1890s. The argument about wood or steel mills went on for a long time, but in time the economy, robustness and efficiency of the steel windmills won. Figure 2.9 shows an Iron Turbine, the first commercially successful all-metal windmill on a ranch in Nebraska in 1886.

One factor that no doubt greatly assisted the optimization and refinement of the American Farm Windpump was the experimental research and development programme of Thomas O. Perry, initiated with the support of the US government's Geological Survey from 1883 onwards (ref. Thomas O Perry, *Experiments with Windmills*, US Geological Survey, Washington DC, 1899). Perry mounted a series of model windmill rotors on a steam-powered roundabout mechanism in a large barn. The rotors could be driven at different speeds to achieve a constant and controlled wind velocity, and the performance characteristics were accurately measured.

This was the first time since Smeaton's experiments a century earlier that serious scientific work was reported on windmill rotors. These studies indicated that the steel, multi-bladed rotor with curved plates for each blade has the best performance over most other options by a good margin. Perry found the maximum achievable rotor efficiency was around 30%; this was discovered empirically and without any real understanding of the aerodynamics involved, but even with modern knowledge it is not possible to improve significantly on multi-blade rotor performance.

During the early 1900s virtually all windpumps were made from steel, self-oiling was introduced and the bearings of the main shaft and the crankshafts were enclosed within special cast-iron reservoirs in an oil bath. At about this time the Aermotor Company (which still exists today) introduced a back-geared windpump where not only was there an oil bath transmission but the rotor drove a small pinion which engaged with a large gear which drove the pump rod via a Pitman

Figure 2.9 The *Iron Turbine* windpump at the Still Gates Ranch, Nebraska, USA, 1886

Crank. This innovation improved the starting characteristics and reduced pump wear and became the standard concept for all but the largest of windpumps from that time until the present.

Large-scale production of American windpumps started in the 1870s and reached its peak between the end of the century and World War I. Many companies went into the business. By the turn of the century the Aermotor company alone had produced over 800,000 windpumps.

The twentieth century

The American Farm Windpump technology spread throughout the world during the 1890s and manufacture of licensed or similar designs took place in Australia, Argentina, South Africa, the UK, France, Italy, Germany, Sweden and the Soviet Union. The windpump export trade reached its peak before World War I, but exports continue to the present day, mainly to South Africa, India, Australia,

New Zealand, Mexico, the West Indies and South America.

In England, the best recorded trial was that held by the Royal Agricultural Society in 1903. It was clear that the preferred type was the fixed-bladed galvanized steel wind-head on top of a steel lattice tower, either three or four sided, with an oil bath containing all the gears and the transfer from rotary to reciprocal motion with an adjustable fantail.

US sales remained healthy into the 1920s. The real problems for the industry became evident in the 1930s. The great depression, gasoline engines and electrification struck the windpump industry. World War II with its government restrictions on production also had a big impact.

The main colonial powers, Britain and France, retained a small market for windpumps immediately after World War II in their colonies, but subsequent independence for most of the colonies led to a further decline in the market for windpumps.

By the late 1950s and 1960s only a handful of makers remained. As energy prices began soaring in the 1970s and there was a revival of interest in renewable energy sources such as wind power, windpumps again began to seem attractive investments. With the renaissance of windmills has emerged a number of interesting innovations.

Wind energy has not only been used for milling and pumping: there were many other uses to which it has been put, such as sawing of timber, polishing of stones, pressing of vegetable oil, production of shingles, and milling of paint, snuff, mustard, lime and many other things. The windmill has also been used as a source of electrical power ever since P. La Cour's mill was built in Denmark in 1890 with patent sails and twin fantails on a steel tower. Wind-electricity was used at the beginning of the century and had a renaissance during World War II, but the real breakthrough came in Denmark and USA in the 1980s after the oil crises.

Today many thousands of large wind turbines are in use in the world for electricity production and the State of California in the USA and Denmark both derive about 2% of their electricity supplies from windmills.

2.2 The Role of Windpumps in Development

In the same way that windpumps were catalytic in opening up to economic development the Great Plains of the USA, the Pategonian Plains of Argentina and the outback of Australia, they are proving to be invaluable in the development of livestock industries in other countries in Africa, Asia and Latin America. In Kenya more than 200 *Kijito* windpumps have been manufactured by Bobs Harries Engineering Limited (BHEL), principally for livestock watering. Other countries with significant livestock herds, such as Botswana, Namibia and Zimababwe, are finding windpumps to be well suited for meeting livestock water demands.

Likewise, as the windpumps were used to drain the lands of the Netherlands and Eastern England, windpumps are being proposed for the large-scale washing and draining of saline flat lands in China, in addition to irrigation and water supply.

India has embarked on a programme of windpump dissemination with full government support and more than 2000 windpumps are now in use in India, mainly for irrigation, but also for water supply and pumping sea water for salt production.

Water pumping is vital for several sectors of development in rural areas of developing countries. Water supply is necessary for the feeding of livestock herds which can be the mainstay of some economies.

For village populations access to clean water is best achieved through pumping from potable water acquifers below ground rather than from using polluted surface-water sources.

With more than half of human sickness in the developing world attributable to poor-quality drinking water, pumped clean water causes a measurable reduction of sickness and health care requirements and leads to increased productive capacity of the population.

Similarly, to increase the productivity of the agricultural sector, irrigation water pumping is necessary if crops are to be grown outside the rainy seasons. Although irrigation pumping using windpumps is less widespread than for drinking-water supply (for reasons covered in more detail in chapter 7) India and Pakistan, for example, have successfully applied indigenous windpumps for irrigation pumping.

Whilst the potential for windpumps is thought to be several million in the developing world, the numbers installed to date remain insignificant in comparison. Only a few developing countries (Table 2.1) manufacture windpumps in significant numbers (more than 100 per year).

Several reasons have been suggested for the slow take-up of windpump technology in developing countries. These are summarized in Table 2.2.

Table 2.1   Developing countries with significant wind-pump manufacture

Africa	Asia/Pacific	LatinAmerica
Kenya	China	Argentina
Mauritania	India	Bolivia
Morocco	Pakistan	Brazil
South Africa	Philippines	Chile
Zimbabwe	Sri Lanka	Peru

Table 2.2   Possible reasons for the slow introduction of windpumps to date in developing countries

- no major international projects for the widespread use of windpumps often resulting from lack of awareness amongst development decision-makers of the advantages of the technology

- poor experiences from some demonstration projects where inappropriate windpumps or applications have been selected or for project infrastructure reasons such as lack of preventative maintenance training or user involvement

- preference amongst some project developers for what are conceived to be more modern technologies such as solar photovoltaic pumps

- preference amongst some development workers for village-level informally-manufactured designs which mainly exhibit poor performance, a short life and poor reliability and are thus considered inappropriate technology

- too much of the development effort for modern windpumps has been expended on complex research to obtain high peformance rather than on achieving ultra-reliability and affordable costs

- inadequate investigation of market requirement

- commercialization activities given lower priorities than research activities by some funding agencies

- an absence of financing mechanisms to allow users to purchase windpumps which although economically competitive may initially cost more than a diesel-driven pumpset

As of early 1992 no major windpump dissemination programmes were being implemented, wheras the installation phase of a US$30 million programme to install 1300 solar photovoltaic pumps in West Africa was underway, supported by the European Development Fund. Other smaller solar pumping projects funded by bilateral agencies are also being implemented in many developing countries, yet the economic competitiveness of windpumps in many regions is widely accepted and is demonstrated in Chapter 7. The European Community is however supporting the development of a new ultra-reliable small diameter windpump under a research contract with IT Power Ltd of the UK and it is hoped that international support for windpumps will increase significantly during the 1990s.

Figure 2.10 A 7.1m diameter *Kijito* windpump manufactured in Kenya by Bobs Harries Engineering Limited, Thika

Figure 2.11 Irrigation windpump in Thailand (without sails)

2.3 Windpump Applications

There are three main applications for windpumps:

- **Water supply** for households, villages and livestock herds. The larger windpumps over 4m diameter are mainly used for community water supply or very high-head pumping. The smaller diameter units are used for household and small herd or low-head livestock watering. Water supply is characterized by a generally constant demand month by month.

- **Irrigation** is characterized by a variable monthly demand with often no water being required in the wet seasons but high volumes required during the planting and dry seasons. Windpumping for irrigation is thus very often less economic than for water supply as the windpump may be underutilized in some months. A match between the irrigation season and wind resource is also needed.

- **Drainage** is the application associated with the development of windmills in the Netherlands for Polder drainage and land reclamation. There is still a requirement for this in many regions around the world. The large volumes often associated with this application adversely effect the economic comparisons with diesel pumpsets or grid extension for electric pumpsets.

- **Minority applications** include pumping of leechate from landfill sites (Europe and the US), water circulation in nature reserves and fish farms and salt water pumping for salt production in coastal regions of India, China and the Mediterranean.

The vast majority of windpumps in use today (around 500,000-750,000) are for livestock and village water supply and to a more limited extent irrigation.

Figure 2.12 Stock watering application for windpumps (*Kijito*, Kenya)

Figure 2.13 Brine pumping for salt production (Cape Verde)

2.4 The Anatomy of a Windpump

The rotor
This can vary widely in both size and design. Diameters range from less than 2m up to 7m. The number of blades can vary from about 6 to 24. In general a rotor with more blades runs slower but is able to pump with more force.

The tower
Normally metal (galvanized steel) with three or four legs. May be anything up to 15m in height, but usually about 10m. The bases of the legs are fixed, often by bolting to concrete foundations.

The pump rod
This transmits the motion from the transmission at the top of the tower to the pump at the bottom of the well. The motion of the pump rod is reciprocating (up and down) and the distance it travels (called the stroke) is typically about 30cm, depending on the pump. Pump rods are usually made of steel.

The well
With a shallow water-table the pump may be mounted on a hand-dug open well; if the level is deep a borehole must be drilled. The outer walls of the well will be lined to prevent in-fill but the liner should be slotted to allow water to enter the well.

The rising main
This is the pipe through which the water is pumped, and also encloses the pump rod.

The tail
Keeps the rotor pointing into the wind, like a weather vane. The whole top assembly pivots on the top of the tower, allowing the rotor to face in any direction. Most machines incorporate a mechanism into the tail which will turn the rotor out of the wind to prevent damage when it becomes too windy.

The transmission
Turns the rotation of the rotor into reciprocating motion (up and down) in the pump rod. Normal types use a gearbox or are direct drive. With direct drive the pump rod moves up and down once for each turn of the rotor. Using a gearbox allows the pump to be geared-down so that it does fewer pumping strokes for a given rotor speed, but with a larger output per stroke.

Well-head components
At the well-head the water is piped away to a storage tank through the 'discharge pipe'. The pump rod usually runs through a seal at the well-head called a 'stuffing box' which prevents water escaping around the rod.

The pump
Normally submerged below water level. On the downward stroke the cylinder fills with water; on the upward stroke the water is lifted by the piston up the riser pipe. Pumps come in various cylinder bores and stroke lengths. The pump hangs on the rising main.

Worldwide Experience and Potential 3

3.1 International Programmes

The IT windpump programme

The origin of the UK IT windpump programme is traceable to the British Engineer Peter Fraenkel who in 1975 was working in Ethiopia on the development of low-cost informally-manufactured windpumps of the Cretan-sail type. The experiences of this work led him to consider that this village level 'do-it-yourself' low cost approach did not result in reliable windpumps and so did not meet the needs of the users.

Thus the Power Project of the Intermediate Technology Development Group (ITDG) of the UK initiated a programme to develop a robust reliable windpump. The windpump design was intended for manufacture in small-scale industrial enterprises in developing countries. The work was taken over from ITDG by an associate company, I T Power in 1981.

In 1977 ITDG identified and began to collaborate with six developing country institutions with an interest in manufacturing windpumps of the direct crank type transmission. These were a non-governmental organisation (the Rural Industries Innovation Centre) in Botswana, private industrial companies in Kenya (Bobs Harries Engineering Ltd), Pakistan (Merin Ltd) and India (Voltas Ltd), the National Research Centre in Egypt and a multinational oil company in Oman.

By 1979, five out of the six collaborators had manufactured and installed working prototypes (Figure 3.1) of the 'Intermediate Technology' or 'IT windpump' (one of which in India was working more than 10 years later - and may still be). The Kenyan collaborator BHEL had gained an order for ten large-diameter windpumps for livestock watering

applications. This forced the pace of development and highlighted the need for a thorough design review. In 1981 I T Power sought and was awarded a research contract by the UK Overseas Development Administration (ODA) to finalize the design of the 6.0m and 7.5m diameter IT windpumps.

Figure 3.1 An early prototype IT windpump in India, 1979

The IT Windpump is of the direct-drive crank-transmission type that requires no casting and thus is easily fabricated in developing countries. The design, which received significant input from the Kenyan collaborator, is now manufactured commercially by original collaborators in Kenya (as the *Kijito* variant) and Pakistan (*Tawana*).

The IT windpump is also manufactured in Nigeria and Zimbabwe (and there are some close copies in other countries). The UK ODA also assisted I T Power in monitoring the performance of installed windpumps and diesel pumps, such that true economic comparisons could be made to verify the economic case for windpumping.

Some 300 plus IT Windpumps have now been installed, mainly in Kenya, and the effectiveness of the design has been recognised by awards from the International Inventors Association of Sweden and the Worshipful Company of Turners of the UK.

From discussions with the IT windpump collaborators and preliminary market assessments it was recognised that a much larger demand existed for small diameter windpumps (<4.0m diameter) than for the existing 6.0 and 7.5m diameter IT windpumps.

Thus in 1990 I T Power embarked on the development of an affordable high performance ultra-reliable design of small windpump with assistance from a European Commission grant. The design has a high performance transmission and will be available in rotor sizes from 1.5 to 4.0m diameter. The first prototype (3.0m diameter) was installed in Spring 1993 (Figure 3.2). The high performance is achieved by the use of a simple quasi-constant torque mechanism.

During 1991/92 I T Power was contracted by the UK Overseas Development Administration to investigate the market for small windpumps. Much of the worldwide experience and market information reported in this chapter was gained through this study.

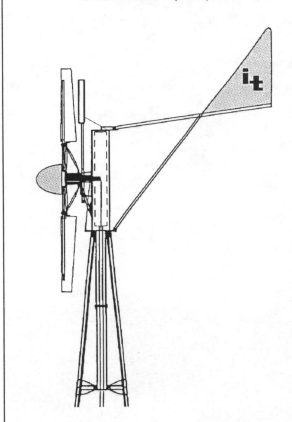

Figure 3.2 A cross-sectional view of the new small diameter (3.0m) IT windpump

The CWD programme

In 1975, at about the same time that the IT windpump programme was beginning, the Dutch established a Steering Committee on Wind Energy for Developing Countries (SWD).

The SWD organization was funded by the Netherlands Ministry for Development Cooperation and was a joint activity of the Eindhoven University of Technology, the Twente University of Technology and DHV consultants. The programme changed its name to Consultancy services in Wind energy for Developing countries (CWD) in the mid 1980s.

CWD developed a range of small diameter windpumps with direct-drive transmission ranging from the 2.0m diameter CWD2000 to 5.0m CWD5000 (see Figure 3.3).

CWD co-operated in the development and manufacture of the CWD range of windpumps in several countries including Cape Verde, Morocco, Mozambique, Nicaragua, Tanzania, Sri Lanka, Sudan, Tunisia and Zambia.

The withdrawal of the support of the Dutch Government came before fully reliable designs

Figure 3.3 Two of the CWD windpumps

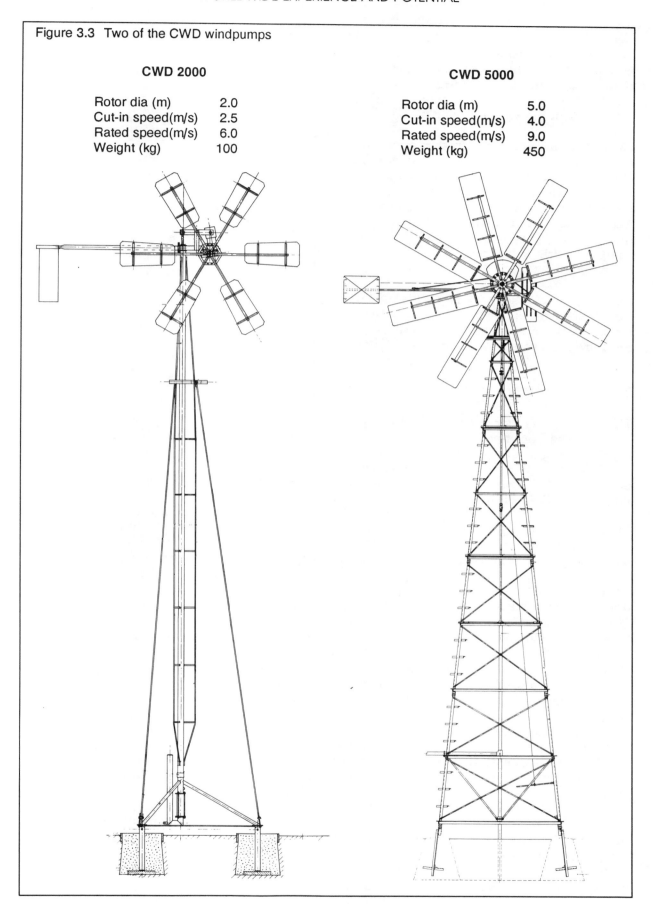

CWD 2000

Rotor dia (m)	2.0
Cut-in speed(m/s)	2.5
Rated speed(m/s)	6.0
Weight (kg)	100

CWD 5000

Rotor dia (m)	5.0
Cut-in speed(m/s)	4.0
Rated speed(m/s)	9.0
Weight (kg)	450

were achieved, and in July 1990 the CWD programme was ended. By that time almost 500 windpumps had been manufactured by their collaborators (i.e. local organizations and workshops in the developing world) but most of them have since ceased production.

CWD concentrated on irrigation applications of windpumps as opposed to water supply. Reasons cited by collaborators for withdrawing from the windpump market included technical problems in adapting the design for manufacture and reliability problems. Economic reasons include windpumps having to compete with subsidized fuel for engine-driven pumpsets.

Windpump activities in the Netherlands are now principally handled by ECN (the Netherlands Energy Research Foundation) and the Eindhoven University of Technology.

The Danish Folkecenter

The Folkecenter for Renewable Energy in Denmark has been working on technologies for developing countries since its foundation in 1983. The first project was the development of a mechanical water pump, in a co-operative project with CWD of the Netherlands.

In co-operation with the Lutheran World Foundation (LWF) a CWD 2740 windpump was installed in Mauritania, and a workshop has been established there for its production. Using the CWD 2740 as a starting point, a 3.6m diameter machine was also developed. This is built in Denmark, but has not yet been installed in any developing countries.

In addition, the so-called 'multi-purpose concept' was developed. This means using an electricity-generating wind turbine with a variety of end-use applications, such as water pumping, grain grinding, lighting, running small workshop machines etc. The turbine (see Figure 3.4) is manufactured in Denmark, and an example has been installed in Zambia. Two other demonstration sites are in preparation, in Gambia and India.

In June 1990 a comprehensive monitoring programme of seven Danish windpumps was completed, six of the machines having mechanical transmissions and one electrical.

Figure 3.4 The *FC4000 Windmotor*, manufactured in Denmark by NM-Electro.

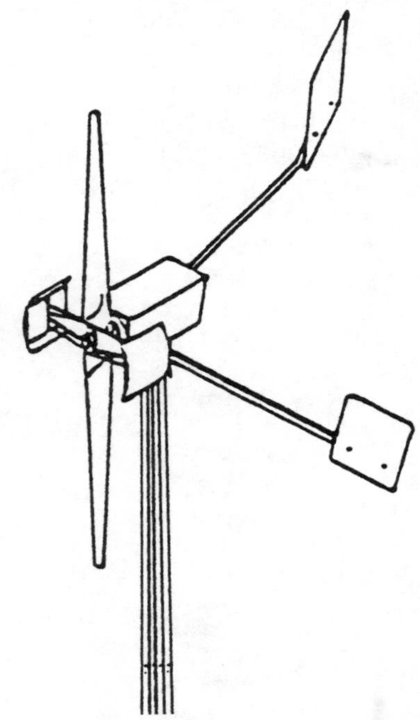

The Global Windpump Evaluation Programme

In November 1986 the World Bank and UNDP initiated a preparatory phase for a Global Windpump Evaluation Programme (GWEP). The preparatory phase included country studies on the current position and prospects for windpumping in seven African, four Asian and three Latin American countries. The preparatory phase also included development of test standards and procedures.

The report of the GWEP preparatory phase concluded that the potential for windpumps remains good and recommended a Global Windpump Implementation Programme (GWIP).

The objectives proposed for GWIP were to have a large windpump installation programme in five regions and to demonstrate that such programmes can be both sustainable and commercial.

Despite the good potential for windpumps that was identified and the stimulus that the programme would have provided, the Global Windpump Implementation Programme was not given the go-ahead.

3.2 Africa

Ethiopia

Ethiopia has been investigating windpumps in a programme of the Ethiopian Science and Technology Commission supported by the Swedish Agency for Research Cooperation (SAREC). Whilst wind speeds are generally low it has been noted that they are greatest during the dry season and that there are areas with annual wind speeds over 2.8m/s and higher on the Red Sea coast suggesting viable windpump application.

Kenya

Kenya has two main manufacturers of windpumps with some other enterprises assembling windpumps informally. There are over 360 windpumps installed, the majority (more than 200) being the *Kijito* windpump based on the IT windpump. The *Kijito* is a large diameter windpump used mainly on deep wells. Over 50% of Kenya has average annual wind speeds over 3m/s. Borehole installations in Kenya average about 170 per year but thousands of shallow wells are in operation.

The existing market for windpumps is approximately 30 units per year, mainly deep-well *Kijito* installations for livestock water supply. There is believed to be a market for smaller diameter windpumps in the coastal, southern and western provinces for small farm holdings and irrigation.

Mozambique

Mozambique was manufacturing approximately 50 units per year to the CWD 2740 design at the Gaza Province Agricultural Department. However production has been interrupted during the recent unrest in the country.

The potential market has been assessed as several thousand by the Department of Agriculture (DPA), particularly for smaller diameter windpumps in the Limpopo and other valleys where wind speeds are reasonable, farm holdings small and diesel pump spares unavailable.

South Africa

South Africa remains the largest market for windpumps in Africa with a reported population in excess of 100,000. New installations have slowed recently as a result of economic problems and displacement by photovoltaic pumps. The market is presently estimated as 800 units per year.

Sudan

Sudan has experience of *Kijito* and Dutch-CWD windpumps installed under UK and Dutch bilateral aid programmes respectively. The *Kijito* windpumps installed to the North of Shendi over-pumped the wells initially and had to be modified to reduce output. Some 150 conventional gearbox design windpumps installed in the 1950s in Gezira have all been abandoned (the last one in 1965) as a result of lack of provision of spare parts and increase in village populations. A 1990/91 USAID in-country study reported that windpumps were appropriate for village water supply but that there would be no significant penetration of the irrigation market in competition with diesel pumps.

Zimbabwe

Two companies, Stewarts & Lloyds and Sheet Metal Kraft (SMK) are manufacturing windpumps. Stewarts & Lloyds are manufacturing the UK designed IT windpump (6.0m diameter) and Climax windpump designs of 2.6m diameter upwards (gearbox types). The SMK design is a 3.6m diameter 18-bladed windpump with ball and roller bearings instead of cast bearings and a fabricated gearbox housing rather than a cast housing. Approximately 200 windpumps have been installed and sales are around 30 per annum.

North Africa

Morocco has a long history of using windpumps and the Centre des Energies Renouvelables (CDER) has been co-operating with CWD on the 5.0m diameter Atlas windpump. One study estimated the market for windpumps in Morocco as between 100 and 200 per year (for water supply). Although wind speeds in Africa are mainly modest, most African countries are considered to have viable windpumping applications.

Tunisia is reported as having a market of 50 per year and has one local manufacturer (SEN) and larger markets are estimated in Algeria, Egypt and Libya.

West Africa

There have been windpump demonstration programmes in Gambia, Senegal, Mali, Niger and Burkina Faso. In Mauritania the French institute IT Dello is developing a programme for the local manufacture of small diameter windpumps to meet the West African market which is estimated (by IT Dello) as over 2000 units. Cape Verde, with strong steady wind speeds, represents a particularly good market for windpumps.

Central and Southern Africa

This area has variable markets for wind-pumps. In addition to South Africa and Zimbabwe, Botswana and Namibia have large livestock industries and have had successful windpumping experiences. Windpumps are manufactured in these countries too. Angola has a programme of renovating derelict wind-pumps. There has been interest from industrial companies in Zambia to manufacture windpumps.

3.3 Asia

China

China has a long history of windpumps with some 200,000 wooden windpumps in use in Jiansu province in 1959. By 1963 the number in use was about 130,000 but the number has declined and is now approximately 1700 in the whole of China.

The Chinese Wind Energy development Centre (CWEDC) is undertaking most development of windpumps, in association with the Chinese Academy of Agricultural Mechanisation Sciences (CAAMS), various

academic institutes and the Xinghua Tractor Factory. High-lift geared windpumps have been developed with rotor diameters from 2.0m (the FD-2) to 6.0m (the FD-6) for use in Inner Mongolia and Gansu.

Low-lift high-volume windpumps for irrigation, drainage and saltwater pumping have also been developed with screw pumps and chain pumps. Figure 3.5 shows windpumps in agricultural use in China.

With good wind resources on the south east coast of China and the plains of Inner Mongolia, and thousands of villages remote from the grid, the potential for windpumps appears high.

Animal husbandry is a major industry in Inner Mongolia for which windpumps for livestock water supply are viable. Some 5000 windpumps are planned for the region over the next ten years.

Figure 3.5 An impressive array of windpumps in use in China

India

Although there are earlier references to windpumps, the first indigenous concerted efforts to apply windpumps in India began in 1952 under the auspices of the Council for Scientific and Industrial Research (CSIR) and later the National Aeronautical Laboratory (NAL). Some 160 Southern Cross windpumps were imported during the late 1950s but most were abandoned after a few years due to lack of trained staff and spare parts. By 1966 some 70 WP-2 locally manufactured machines were fabricated but then production ceased through lack of interest. After the 1973 oil crisis, interest was revived by the Department of Science and Technology and a national demonstration programme initiated for the sixth plan (1980-85).

By the end of the sixth plan more than 1700 windpumps had been installed, mainly of the 12-PU-500 model manufactured by research institutes. The original 12-PU-500 'appropriate technology' windpump developed by the organization of the Rural Poor in Ghaziapur and the Dutch TOOL Foundation proved to be unreliable.

Recognising the problems experienced with the 12-PU-500 windpump, the wind programme during the 7th plan (1985-90) concentrated on the introduction of new designs of windpumps, and institutional strengthening to assist installation, preventative maintenance and repair. The commercialization of windpumps accelerated and several commercial manufacturers introduced new designs and improved versions of the 12-PU-500. By the end of 1992 some 3500 windpumps had been installed and the Government's Eighth Plan (1990-95) calls for a target of 5000 windpumps installed with full commercialization.

There are approximately 20 companies manufacturing or assembling windpumps in India. The most active of the windpump manufacturers are NEPC in Madras, BHEL in Rudrapur, the Scientific Instrument Company in Madras and Wind Fab in Coimbatore. NEPC market a conventional gearbox transmission windpump known as the Green Rev (see Figure 3.6). BHEL have developed an indigenous direct-drive (crank) windpump and have been undertaking comparative tests with an

Figure 3.6 The NEPC *Green Rev* windpump (installed in Madras).

imported FIASA windpump. The Scientific Instrument Company have a joint venture with WD Moore of Australia for the manufacture of the Yellowtail windpump. Wind Fab market a 2.5 and 3.0m diameter Blue Vane windpump with gearbox transmission.

The 5.0m diameter 12-PU-500 machine is manufactured by several companies (with varying design adaptations and reliability). Typical manufacturing costs are Rs15,000 (US$250) of which 80% is accounted for as material costs.

The conventional gearbox windpumps suitable for deep well installations are sold at typically Rs45,000 (US$1600). This is four times more expensive than a diesel pumpset and thus is the principal barrier to widespread use despite being economic on a life-cycle cost basis. The vast majority (80+%) of windpumps installed in India are of the 5.0m diameter 12-PU-500 type applied principally to low-head irrigation. However, this is not indicative of the market requirements but of the target application of the first National Demonstration Programme.

Most manufacturers are moving towards other windpump designs capable of higher-head pumping. Windpumps are also used in the pumping of sea water for salt production in Gujurat and some other states. The wind resource in India is favourable in some states such as Tamil Nadu, Gujurat and regions of Andhra Pradesh but not in all states.

Various studies have been undertaken to estimate the potential for windpumps in India including those that were part of the World Bank supported Global Windpump Evaluation Programme. The potential has been estimated as approximately 100,000 units.

Indonesia

Indonesia has an ongoing research programme organised by the National Institute for Aeronautics and Space (LAPAN) with indigenous designs (SM-03) and (SM-05) in production. Other indigenous Indonesian windpumps are used in the salt industry on the North coast of Java.

Mongolia

Mongolia has imported Russian windpumps for the livestock industry but was reportedly dissatisfied with the technology. Small diameter Akrobaatti windpumps from Finland were also demonstrated under a UNDP programme. Given the 100,000 nomadic livestock-herding families in Mongolia, a potential market of several thousand windpumps is assumed for livestock-watering stations.

Pakistan

Pakistan has the potential for windpumps in the coastal regions, particularly in Sind Province. The local manufacturer, Merin Limited, manufactures the Tawana windpump (Figure 3.7) which is based on the IT Windpump (6.0m and 7.5m diameter). The company has exported its Tawana windpumps to Nigeria and the Middle East. A commercial market of 200 to 300 windpumps per year is estimated.

The Philippines

The Philippines has a high potential for windpumps, principally for water supply. One study assessed the potential as more than 600,000 based on the number of Local Utility Administrations that have not set up water supply systems. More than 200 conventional gearbox windpumps have been manufactured and installed locally and many are reportedly working successfully after more than ten years operation. There have also been various Savonius rotor windpumps developed in the Philippines although none are considered to have commercial potential. The Philippines National Construction Company has used windpumps for water supply applications.

Sri Lanka

Sri Lanka has had a poor experience with windpumps as the locally manufactured WEU I/3 windpump suffered from reliability problems. The Government Water Resource Board terminated its windpump programme in 1987. The microeconomic conditions for small (2-4m) diameter windpumps remain good however, and if a reliable windpump can be introduced, a commercial market of more than 150 units per year is estimated. In 1983 the Dutch windpump agency, CWD, estimated that 10,000 of the 50,000 diesel pumps in Sri Lanka could be economically replaced with windpumps.

The University of Moratuwa and other organizations are keen to reactivate the development and manufacture of reliable windpumps. One company in Anuradhapura continues with the manufacture of an improved design of the 3.0m diameter windpump.

Thailand

Thailand has a windpump industry that has seen production of more than 2500 units. Research is being undertaken by King Mongkuts Institute of Technology.

In other parts of Asia (including Vietnam, Malaysia, Laos and Cambodia) windpumps are being applied and some manufactured locally. Resistance to typhoons in some regions remains a problem.

Figure 3.7 Erecting a *Tawana* windpump manufactured by Merin Ltd, Pakistan

3.4 Europe and the CIS

There are only a few manufacturers of windpumps left in Europe, notably Abachem (UK), Tozzi and Bardi (Italy, see Figure 3.8) and Poncelet 'Oasis' (France).

In Eastern Europe a few manufacturers also exist. The market for windpumps in Europe is only a hundred or so per year with Greece and Turkey the main markets.

The CIS, which reaches from Western Europe to the East of Asia, has a vast potential windpump market and indigenous designs (developed by the All Union Institute for the Electrification of Agriculture, VIESH) have been developed and marketed. The principal markets are in Uzbekistan, Turkmenistan, Khazakstan, Kyrgyzstan and Eastern Russia.

A recent assessment by the Wind Power Association of Russia reported a potential market of more than 500,000. One of the factors reported is an increasing need to take water from ground sources rather than polluted surface-water sources.

Figure 3.8 A *Tozzi & Bardi* (Italy) windpump, installed at Matola, Mozambique

3.5 South America and the Caribbean

Argentina

Argentina is one of the success stories of windpumping in developing countries. Output is around 5000 per year representing a significant proportion of global windpump production.

In 1876 the first windpumps were imported from the USA and subsequently manufactured by the Lanus Roland company. Large scale manufacture of windpumps was started by the Instilar Co. in 1937. In the 1960s the Aermotor windpump company of the USA granted a licence to a company manufacturing agricultural equipment (including FIAT engines under licence).

The subsequent production of this 1930s-design gearbox windpump to satisfy the US and the Argentinian markets (principally for livestock watering) soon grew and the FIASA windpump is now one of the most widely-sold windpumps. There are also at least 10 other manufacturers of windpumps in Argentina.

The total windpump population is reported at between 300,000 and 600,000 and it has been estimated that approximately 65% of these are with rotor diameters less than 10' (3.4m). The price of Argentinian windpumps are typically US$500 for an 8' (2.7m) diameter machine and US$800 for a 10' (3.4m) diameter machine ex-works. Wind speeds in Patagonia average 4.5m/s, and in the wet pampa region around 2.5m/s.

Various reasons have been offered for the success of windpumps in Argentina which has been achieved without any government involvement or assistance and few government orders. These factors can be summarized as:

- high wind speeds (compared to Africa for example)

- large livestock industry

- initial guaranteed export market (to US licenser)

- sub-contracted component manufacture involving large numbers of companies and thus many experienced people

- sub-contracted maintenance and repair, letting manufacturers concentrate on production and marketing

- the development of teams of 'Jagueleros' to undertake major repairs and servicing

- good user maintenance

Bolivia

Bolivia has been producing the *Condor* 2.0m diameter simple windpump based on the Gaviotas design of Colombia. Two other organizations, CIPER and SEMTA, are producing designs based on the TOOL 12-PU-350 and 12-PU-500 designs.

Brazil

Brazil has 12 manufacturers producing conventional gearbox windpumps with some 7000 or more installed and almost all under 4.0m diameter.

A particular market is seen as Rio Grande do Sul and the livestock-ranching areas. Lack of disposable income of livestock ranchers in recent years has slowed the market and some inroads have been made into the market by the indigenous photovoltaic pump manufacturer Heliodinamica. The University of Pernambuco in Recife believe non-vertical boreholes and insufficient wind in the arid areas makes pumps remote from the windturbine a viable technology in Brazil. In general a potential market for small windpumps of around 700 per year has been estimated.

Colombia

Colombia, after Argentina, is South America's most significant windpump market with with some 3000 imported units installed (of which half are abandoned) and a thriving local industry (Las Gaviotas) which has sold 9000 MV2E 2.0m diameter machines with an

Figure 3.9 One of the windpump designs of *Gaviotas* of Colombia

The rest of South America

In addition to Argentina, Brazil, Colombia and Peru, the South American countries Chile, Ecuador and Venezuela are known to have significant potential markets but few details are available on the existing market.

Central America, Mexico and the Caribbean

Mexico has imported windpumps principally from the USA. In Nicaragua IMEP manufacture a 3.0m diameter conventional gearbox windpump at a production rate of approximately 80 per year.

Several Caribbean countries have installed windpumps but there is little indigenous industry.

3.6 Rest of the World

Australia

Some 150,000 windpumps are in use in Australia but production peaked at 10,000 per year in 1965 and has declined since, due to reducing livestock numbers and new alternatives, including solar pumps.

The Yellowtail, manufactured by WD Moore, is one of the more successful small diameter gearbox windpumps with 6, 8, 10, 12 and 14 foot sizes. Southern Cross are the other principal manufacturer with windpumps from 6 to 25 feet in diameter.

The Australian market is now estimated at around 500 units per year.

annual market of over 1000 units (see figure 3.9). These machines are limited to low-head pumping and so the company has developed a higher head (up to 50m) 3.1m diameter *Guajira* windpump.

Peru

Peru is one of the few areas where indigenous appropriate technology type windpumps are in use in significant numbers. More than 1000 of the *Miramar* 6m diameter wooden units are in use on the Pacific coast.

There are eight other windpump manufacturers in Peru all but one of which (Matto) manufactures windpumps under 4.5m diameter. Of the metal fabricated machines the lowest cost unit is the Segovia 2.8m diameter gearbox type windpump selling at approximately US$800.

The potential windpump market in Peru is estimated at approximately 600 units per year.

USA

The USA has also seen a decline in windpump sales as a result of increasing electrification and sales of photovoltaic pumps.

Although there are still several active manufacturers (e.g. Aermotor, Dempster) the annual market is now believed to be less than 500 units per year.

3.7 Summary of the Market Potential

An estimate of the existing and potential markets for windpumps is given in Table 3.1 and also shown as a pie chart in Figure 3.11. The potential market is an estimate of demand, given reasonable promotion and market development resources.

It can be seen that Asia, Latin America and the CIS are considered the main markets. Europe is considered a relatively small market and the USA, Australia and South Africa declining markets (principally because of a downturn in the livestock industry, further rural electrification and use of photovoltaic pumps). Photovoltaic pumps are not readily available without hard currency and significant photovoltaic manufacture is present in only a few developing countries (e.g. Brazil, India and China).

It is expected that the windpump market will grow by a factor of five, whilst the wind-pump population will mostly remain constant. This is because there will be a continuing fall in the windpump population in Australia, USA and South Africa, but new markets will develop in other countries.

The applications considered to have most potential are livestock watering and village and homestead drinking-water supply.

Figure 3.10 Photovoltaic pumps displacing windpumps in the USA

Table 3.1 The potential market for windpumps				
Region	Existing Market (units)		Potential Market (units)	
	Total	Annual	Total	Annual
India	4,000	500	100,000	6,000
Rest of Asia	4,000	500	200,000	12,000
South Africa	200,000	800	55,000	5,000
Rest of Africa	3,000	400	15,000	2,000
Latin America	350,000	7,000	400,000	20,000
Europe	5,000	300	10,000	1,000
CIS	8,000	500	100,000	10,000
North America	50,000	500	10,000	1,000
Australia/Pacific	150,000	800	10,000	1,000
Totals	770,000+	12,000+	900,000	58,000

Figure 3.11 Breakdown of the existing and potential windpump market

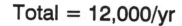

Total = 12,000/yr

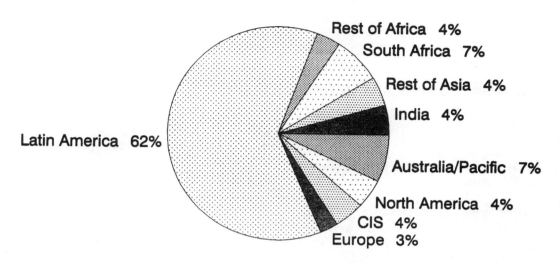

Rest of Africa 4%
South Africa 7%
Rest of Asia 4%
India 4%
Australia/Pacific 7%
North America 4%
CIS 4%
Europe 3%
Latin America 62%

**Existing market
(Units)**

Total = 58,000/yr

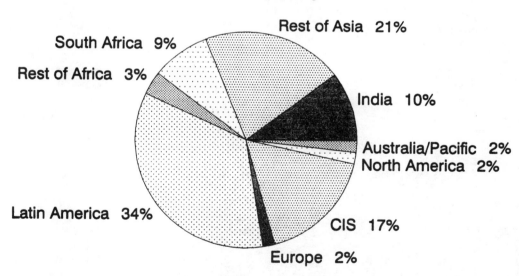

Rest of Asia 21%
South Africa 9%
Rest of Africa 3%
India 10%
Australia/Pacific 2%
North America 2%
Latin America 34%
CIS 17%
Europe 2%

**Potential market
(Units)**

The Wind Resource 4

4.1 Energy from the Wind

To understand how the wind can be best used, it is necessary to have at least a qualitative understanding of the nature of the energy in the wind, and the things that affect the wind's behaviour.

This chapter looks at the ways that the resource can vary with both time and location. It also provides simple rules to help the potential installer assess the suitability of their site.

The wind contains energy by virtue of its motion, and this is called kinetic energy. A windpump makes use of this energy by slowing the wind. To remove all the energy from the wind it would be necessary to bring it to a complete stop, but this is not theoretically possible. It has been proven that the best achievable efficiency is when the wind is slowed by the turbine to one third of its upstream velocity, where 60% of its total energy can be extracted.

A real windpump will only be able to capture about 30 of the total energy in the wind. There will also be energy losses in the transmission and in the pump itself, whose combined efficiency will be about 70%.

Lifting a certain volume of water per second through a certain head requires work and this can be expressed as a 'power' (measured in watts). Overall there are three important points that relate to the power produced by a windpump:

- The theoretical power in the wind is proportional to the cube of the windspeed. This means that if the windspeed is doubled the power increases by a factor of eight. Clearly this has important implications for siting, as a small increase in mean windspeed will pay large dividends in terms of the extra power obtainable.

- The output power is proportional to the area swept by the rotor blades. It is therefore proportional to the square of the rotor diameter.

- The power in the wind is also proportional to the density of the air: a certain volume of air will have a greater mass and therefore more momentum at higher density. At high altitudes the air density is lower and so less power can be obtained for the same windspeed and rotor size.

Figure 4.1 The power of the wind

Altitude

Some remote parts of the world where wind-pumps may be situated are at a high altitude. For instance the Altiplano (literally 'high plain') of Bolivia has its base level around 4000 metres. As described above, the lower air density at high altitudes will reduce the available power for a given windspeed. The power produced at a certain windspeed is reduced by about 10% for each 1000m (3300ft) above sea level.

Variation with time

To assess the case for a windpump it is important to consider not only the mean windspeed, but also the way the mean is made up over a period of time. Perhaps the crudest statistic one is likely to encounter is the annual mean windspeed. This will give a first-guess idea of the viability of the site, but is by no means a sufficient criterion. Although the variation in the annual mean from year to year may be fairly small, the seasonal variation within the year is likely to be large.

A more useful statistic is the mean monthly windspeed. This will take account of seasonal changes and is the quantity most often used for siting studies. However, it should be realized that the fluctuation in mean monthly windspeed for a certain month between different years can be quite large.

For periods of less than a month there is too much fluctuation for mean statistics to be of great use, but it is still necessary to know the way that the wind will change. Of particular relevance is the number of consecutive wind-less or low-wind days in each month. It is important to size the windpump and the storage tank so that there is sufficient water to last over the longest likely lull period.

Figure 4.2 shows variations of the mean windspeeds on three different timescales. This variation consists of cyclic patterns associated with the passing of the seasons, and also of random variations due to the unpredictable nature of the climate. The shorter the mean period that is taken, the larger the fluctuations about that mean that should be expected. The size and distribution of these variations will depend on the type of climate that influences the region.

Diurnal effects

An important type of cyclic time variation that will be almost universally experienced is the change in windspeed between day and night. During the night, the wind at the surface tends to be less strong than during the day, although the wind higher up will be the same

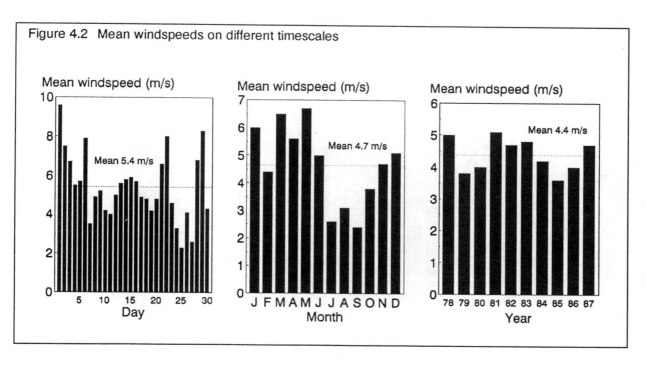

Figure 4.2 Mean windspeeds on different timescales

regardless of the time of day. This is an indirect effect of sunlight heating and cooling the ground. At night when the ground becomes cool it also cools the lower layers of the air. This has the effect they become 'separated' from the air higher up, and tend not to move with the upper wind. This will only affect the lower 50m or so of the atmosphere, but the depth of surface air that is separated grows upwards frm the ground during the night.

The separation effect is most pronounced on clear nights when the ground cools very quickly. During the day the heating of the lower layers soon reverses the phenomena, and the wind extends down to the surface again. Figure 4.3 shows an example of such a diurnal cycle.

The mean power in the wind

Some care must be exercised when using mean windspeed data, as the same wind-pump on two different sites which have the same mean windspeed will not necessarily produce the same power output. This is because the mean power (and so the mean quantity of water lifted) over a period of time depends on the windspeed distribution; in other words, on the frequency of occurrence of different windspeeds.

The reason for this is that the mean power depends on the mean of the cubes of the windspeeds over that time, and this is different to the cube of the mean windspeed. This can be a confusing concept, but in general it can be said that the actual power produced will be larger than that calculated from the mean windspeed. This is because a disproportionately large amount of power will be produced at those times when the windspeed is higher compared to when the winds are lighter.

A graph illustrating the frequency distribution of windspeed for a typical site is shown in Figure 4.4. The horizontal axis shows the range of possible windspeeds and the vertical axis shows how frequently they occur. This shape can be approximated by a statistical curve called a Weibull distribution. Depending on the shape of this curve, the real power available will be greater than that calculated from the mean windspeed by a factor ranging anywhere from 1.2 to 4. This factor is called the energy pattern factor, and for most localities it has a value of about 2.

Actual mean power = 2 x (power expected from mean windspeed)

The method used in later sections to calculate the windpump output at different

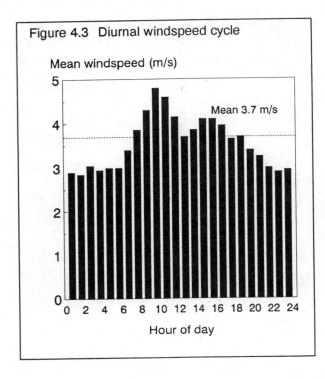

Figure 4.3 Diurnal windspeed cycle

Mean windspeed (m/s)

Mean 3.7 m/s

Hour of day

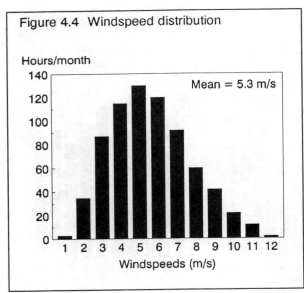

Figure 4.4 Windspeed distribution

Hours/month

Mean = 5.3 m/s

Windspeeds (m/s)

windspeeds takes account of the likely windspeed distribution, and assumes an energy pattern factor of 2. This makes it possible to work with the mean windspeed, and still get a realistic answer for most sites.

Turbulence

When the wind is blowing smoothly and steadily the air-flow is said to be laminar. This means that the air-stream moves in a smooth uninterrupted sheet. This is ideal for windpumps, because the blades need a smooth flow over them to work properly.

However, if the air-flow is suddenly disturbed this can give rise to turbulence. This means that within the overall flow some parcels of air will be overturning and swirling in different directions.

This kind of air flow will disrupt the smooth flow over the blades and reduce the efficiency of the pump. Because turbulent motion is random, some parts of the air stream will move faster than others, and this can put unequal and rapidly changing stresses on the windpump rotor. Although windpumps are designed to cope with this, siting in a very turbulent location will tend to shorten the life of the pump and increase the need for maintenance. It should also be realized that the strength of brief individual gusts may far exceed the mean windspeed.

Therefore when siting a windpump, features in the landscape that produce turbulence should be avoided. These include trees, buildings and certain forms of topography. The effects of these features and some rules of thumb for siting near them are given later in this chapter.

4.2 Local Effects

The wind resource is extremely variable on very small scales of distance right down to a few tens of metres. So even though the prevailing wind over a region may be very good this does not mean that a particular site will have a reliable supply of wind. Conversely, there may be local effects that can generate a good wind resource, even if the prevailing wind is very poor. These factors mean that siting a windpump effectively can be very difficult, and many factors need to be considered. This section aims to give a basic understanding of the more common local effects, and to give some rules of thumb concerning them.

Surface roughness

It is the general case that the wind is stronger higher up, and weaker near the ground. This is due simply to friction between the wind and ground. Put another way, the roughness of the surface slows the wind down. Higher up, the windspeed is closer to that of the prevailing wind. This effect is most pronounced in the first few tens of metres above the surface, and so the higher the windpump tower, the better the mean wind. Because the output power is proportional to the cube of the windspeed even a small increase in mean speed will give a sizeable increase in power. Using a larger tower to help overcome the effect of surface roughness is one of the most cost effective ways to increase the effective wind resource.

The type of surface affects the way that the wind is slowed, and the height above the surface for which it is affected. For instance, trees will offer more resistance to the wind than grass, and so the wind will be slowed to a greater altitude above them.

To aid in decisions on tower height, it is useful to know how the windspeed varies with height over different surfaces. This can be visualized with wind profiles such as the example shown in Figure 4.5. The horizontal axis of the profile represents the windspeed. The wind high up is taken to be 100%. It can be seen that over a flat surface such as water the speed increases very quickly with height, reaching 70% at only about 25m. But over trees or buildings, the increase of speed is much more gradual.

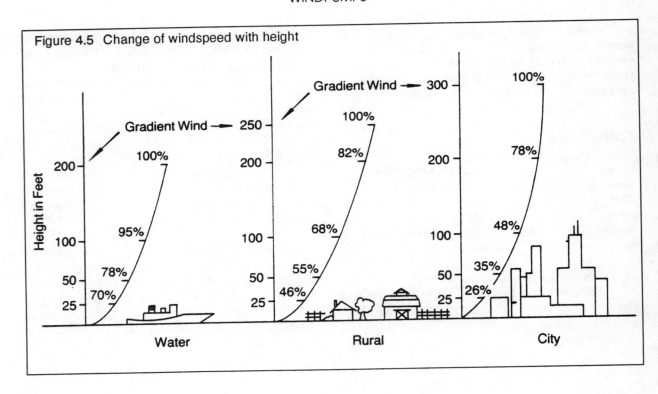

Figure 4.5 Change of windspeed with height

However, it is probable that measurements will only exist for surface windspeed and the windspeed very high up will not be known. Figure 4.6 shows a more general set of wind profiles for different types of surfaces. The figure is constructed with 10m as the reference height, as this is the standard height at which meteorological data are usually taken. Therefore if the wind is known at 10m it can be calculated at any other height. Each of the curves represents a different surface type, which have been arbitrarily divided up as shown in the key, in classes ranging from flat snow to town suburbs.

To use the graph, choose a height at which you want to know the windspeed and select the appropriate curve for your surface type. Trace a horizontal line across from the left hand axis at the height (in m) that you want, to where it intersects the relevant curve. Then trace a line vertically down to the bottom axis and read off the windspeed multiplier. Multiply this number by the windspeed at 10m to find the windspeed at the height you selected.

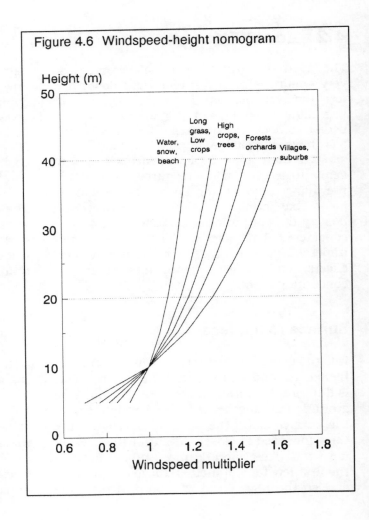

Figure 4.6 Windspeed-height nomogram

In a region with different surface types around the windpump, it may also be necessary to consider from which direction the wind usually blows. Clearly, the best results will be obtained if the windpump is positioned such that there is a smoother surface in the normal upwind direction. Grass is always preferable to shrubs, trees or buildings. Where there is a change in surface type, the characteristic wind profile changes its shape over a horizontal distance of about 50 to 100m.

A final consideration is that there is a small variation of windspeed across the diameter of the rotor. A large rotor positioned too near the ground may be in the surface region where the windspeed is changing rapidly with height: this will mean that there are unequal stresses across the diameter of the rotor which may lead to damage and premature failure.

Effect of obstacles upwind

If there are obstacles upwind of the windpump they will interrupt the air flow and cause turbulence in their wake. This turbulence will extend some considerable distance downwind of the obstacle, and a windpump placed in this turbulent flow will not work at its full potential. It is therefore important to site windpumps such that interference from possible obstructions, such as buildings or trees is minimal.

Figures 4.7(a) and (b) show the extent of turbulence around groups of isolated trees and buildings respectively. Note that the turbulent region extends downwind a distance of about 15 to 20 times the height of the obstruction. The effect is more severe for buildings because they present a sharper barrier to the wind and have a higher solidity. Note also that there is a smaller, but still significant, region of turbulence upwind of the obstacle extending to between about 2 and 5 times its height.

It can be seen that turbulent envelope extends to a smaller height as the downwind distance increases. If siting is a problem due to obstacles then it may be possible to erect the windpump with a tower large enough to raise the rotor above the turbulent region. If this approach is used it should be remembered that the turbulent area also extends upwards to a maximum height of about double the obstacle height.

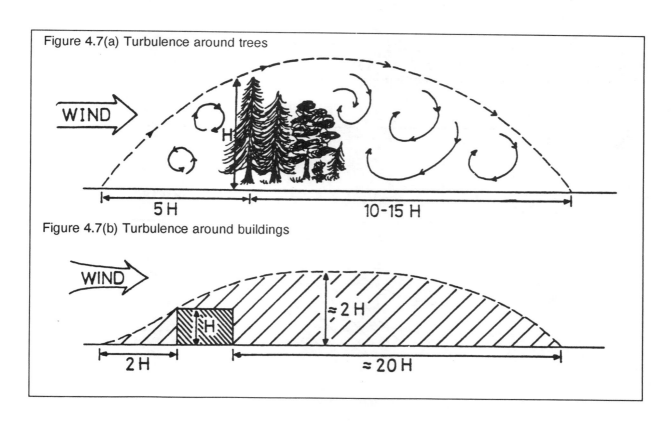

Figure 4.7(a) Turbulence around trees

WIND

5 H   10-15 H

Figure 4.7(b) Turbulence around buildings

WIND

≈ 2 H

2 H   ≈ 20 H

If obstacles are not isolated, but are spaced within range of each others' wakes (e.g. woodland, towns) then the main air-flow tends to lift off the ground. In that case the situation is more similar to the surface roughness effect dealt with above.

Clearly the ideal situation would be to place the windpump well away from all conceivable obstructions, but in many cases this will not be possible. The windpump should therefore be sited such that it is clear of obstacles' wakes in the normal upwind direction. For instance if the wind is almost always from the north, then a building fairly close by, but due east of the windpump should not cause problems.

Topography

Windpumps are most commonly used on flat plains, and not in hilly country. The wind regime is likely to be far more erratic in hills or mountains, and very complex wind patterns can develop. Many of the effects of topography are intuitive: for instance, the wind can only blow along valleys, and the tops of hills are windier than the bottoms.

For the case of hills and ridges, there are some rules of thumb that can be used. A general case that can be applied to most situations is illustrated in Figure 4.8. This shows the way that as the wind is lifted over the hill it is accelerated near the top to about twice the original surface windspeed. But also notice that at the foot of the hill on either side, there is a calmer region where the windspeed is only half of that further from the hill. If the downwind side of the hill is too steep, there may also be an area of turbulence generated at the foot, as illustrated in Figure 4.8.

Recent research has shown that the speed up over an isolated hill is in general greater than that over a ridge. However, in practice it is unlikely that water will be found on hill tops, and so perhaps the main lesson is to avoid the feet of hills (both upwind and downwind) where speeds will be low and the flow possibly turbulent.

Valleys will usually have lower and more variable windspeeds than flat ground or surrounding hills, and are unlikely to be suitable windpump sites, unless aligned with the prevailing wind

Coastal effects - sea breeze

Coastal areas (and lake shores) will tend to have higher windspeeds than inland areas when the wind comes across the water. This is due to the very low surface roughness as described above.

It is therefore important to know from which directions the wind usually blows to decide if a coastal location has any benefit.

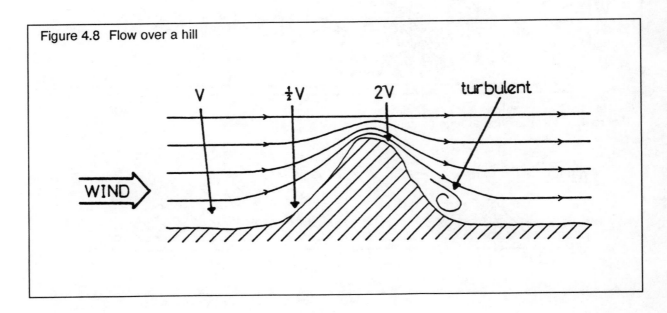

Figure 4.8 Flow over a hill

Another local coastal effect is that of the sea breeze. This is caused by the heating of the land during the daytime, so that it becomes warmer than the sea. The hot air rises and the cool air from the sea moves inland to take its place. This effect reaches its maximum in the mid-afternoon and will continue until around dusk.

At night as the land cools the opposite happens, and the wind blows off the land (called a land breeze). This tends to be weaker than the daytime sea breeze. Figure 4.9 illustrates the sea breeze effect.

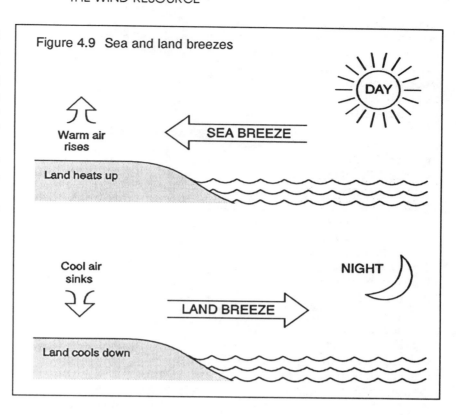

Figure 4.9 Sea and land breezes

Mountain slope winds

In wide valleys or flat regions close to mountains there are mountain breezes produced by the heating and cooling of the mountain slopes. During the day the sun warms the slopes and the warm air rises up them, while cooler air settles in the valleys. This up-slope wind is called an anabatic wind.

At night the situation is reversed and as the ground cools, it cools the air on the slopes.

This cold air then runs down the slope towards the valleys, creating a down-slope wind. This is called a katabatic wind. They may also be caused by lying snow cooling the air on the upper slopes, or a cold air-mass flowing down off a high plateau.

If there is a strong prevailing wind, it may disrupt the mountain wind circulation, and the above effects will not be noticed.

4.3 The Global Wind Resource

The earth's atmosphere has a large-scale pattern of circulation that is ultimately driven by temperature differences between the tropics and the poles, and also by the temperature differences between the continents and the oceans. Although there are variations in this pattern, the annual cycle remains basically the same for a given location. Therefore each area will have a well-known wind climatology, in which the overall prevailing wind blows from various directions at different times of year. Examples of such large-scale

circulations are the westerly trade winds over the UK, or the annual monsoon cycle across southern Asia.

It is possible to make some generalizations about the the global wind pattern, and these apply particularly well in coastal areas or on islands where the effects of land do not disturb the global circulation. Around the equator is a narrow belt of calm air and very light winds called the doldrums, whose position and width varies with the seasons. North and south from the equator the prevailing winds are fairly constant in strength, and blow from the northeast and southeast respectively. These are called the easterly trade winds. Around 30° north and south is another area of lighter winds, and further poleward the winds are westerly. The area around 50 to 60° north or south is characterized by frequent storms and marks the position of the 'polar front' between the cold polar easterlies and the warmer westerlies. This pattern of circulation is illustrated in Figure 4.10.

The local effects that were described in Section 4.2 act in addition to the large-scale circulation, enhancing, reducing or complementing the prevailing wind.

The box below shows the developing countries that have a mean annual wind regime in at least part of their territories of over about 4m/s (countries with population less than 25,000 are not included). It must be realized that the wind regime will still vary greatly from site to site within these countries, and

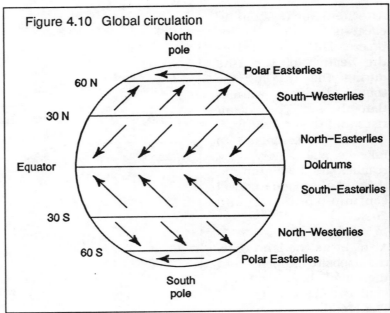

Figure 4.10 Global circulation

that each situation still needs to be assessed on its own merits.

It is perhaps more useful to look at the wind resource on a regional basis. The maps shown in Figures 4.12 to 4.14 illustrate the winds across South America, Africa and Asia. These show the direction and areas of influence of the various winds in more detail, and the text accompanying each section gives a summary of the wind climatology for that region.

Developing countries with best wind potential
Annual mean windspeed greater than 4.2m/s in some part of territory

Africa	Asia	Latin America
Cape Verde	China	Argentina
Chad	India	Brazil
Djibouti	Jordan	Chile
Egypt	Korea	Columbia
Ethiopia	Korean Dem.	Haiti
Kenya	Lebanon	Mexico
Mali	Mongolia	Peru
Mauritania	Sri Lanka	Uruguay
Morocco	Syria	
Namibia	Turkey	
Niger	Yemen	
Senegal	Yemen Dem.	
Somalia		
Sudan		
Tanzania		
Tunisia		

South America

A map showing the principal winds affecting the South American continent is shown in Figure 4.11. The northern countries of Venezuela, Guyana, Surinam, French Guiana and the northern area of Brazil are all north of the equator and experience the northeasterly trade winds common to those latitudes. Further south in Brazil, the prevailing wind is due to the corresponding southeasterly trade winds. Both of these easterlies push a great distance inland over the low-lying Amazon Basin, across as far as the Andes mountain range.

The Andes run in a chain down though Peru and Bolivia and into Argentina. They act as an effective barrier to the wind, separating the eastern and western wind regimes of the continent. The southern part of the continent is subject to the westerly trades. In the far south of Chile and Argentina the land mass is lower and narrow and does not greatly deflect the wind. The southern tip of Chile is one of the windiest places on Earth, with mean speeds around 13 m/s. Further north on the Pacific coast the Andes become more prominent and deflect the westerlies northwards along the coast. These then curve back on themselves to become southeasterlies over Peru and Equador.

The west coast is prone to very strong land and sea breezes, and there are various local mountain winds associated with the Andes.

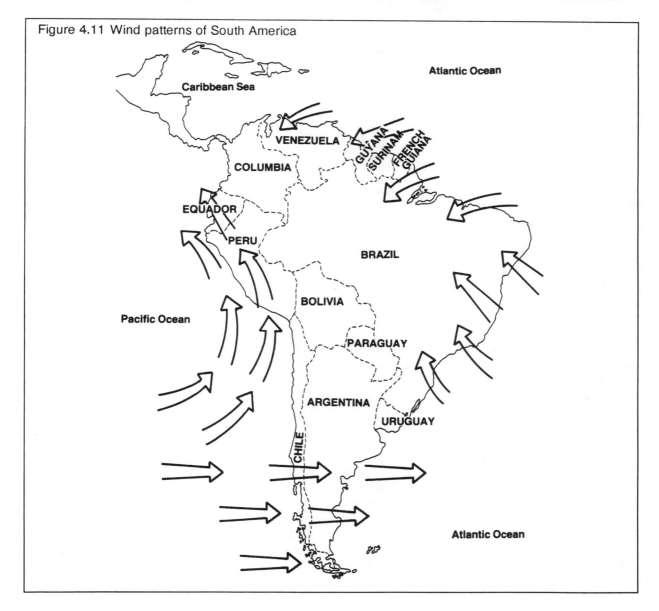

Figure 4.11 Wind patterns of South America

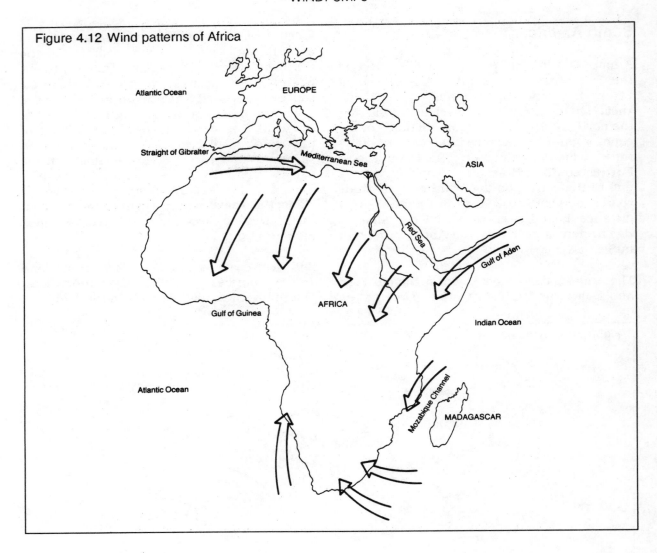

Figure 4.12 Wind patterns of Africa

Africa

A map showing the principal winds affecting the African continent is shown in Figure 4.12. The northernmost coastal area is influenced by the Mediterranean and experiences westerlies. From November to March the winds over most of northern and central Africa are from the northeast (the Harmattan).

In the western bulge these take the form of northeasterly trade winds. Across the interior the direction becomes more northerly, and in the eastern coastal regions the wind is a northeasterly from Asia and the Indian Ocean. This extends some way down the east coast to the Mozambique channel.

The countries on the eastern side of the southern tip of the continent are exposed to the southeast trades from the Indian Ocean. These become deflected around the coast on the west side of the Cape and turn inland becoming southwesterly.

During the period May to September the whole pattern moves north. The northerly winds recede and the southeast and southwesterlies push far inland. During the transition months of April and October winds tend to be light in most of Africa.

Africa does not have large mountain barriers, but it does have vast high plateaus at around 1500m. These will tend to affect local wind patterns, particularly in East Africa. In general the winds of Africa are fairly predictable throughout the year, blowing steadily during daylight hours and dropping off to near calm at night.

Asia

Figure 4.13 shows the principal winds that blow across Asia. The northern part of the continent (Russia and the CIS) is dominated by cold north and northwesterlies bringing air from the polar regions. These blow steadily across the interior. It is only the peninsulas of India and Southeast Asia that extend south from the main continent and experience maritime effects.

The annual pattern of wind in India is dominated by the monsoon circulation which effectively reverses the prevailing wind between winter and summer. A simplified explanation is as follows: during the winter the inner continent cools and cold dense air rushes out over the warm Indian Ocean to form the northeast monsoon. During the summer, the land mass heats up and the rising warm air causes an in-flow of moist air from the Ocean, bringing the wet southwest monsoon. In this respect it is rather like a sea breeze effect. The Himalaya tend to shelter India from the strong winter northeasterlies. They also deflect the summer southwesterlies back on themselves up the Ganges valley becoming more southeasterly towards the Punjab. In general Indian winds are light to moderate with a steady onshore breeze.

Southeast Asia also experiences the seasonal monsoon reversal. Unlike India, Southeast Asia has no mountain barrier, and so the northeasterly winter winds from the continent are stronger. The eastern area is also affected by the northeast trades from the Pacific. During the transition period around April and October winds are light and variable.

In the islands off the south of the peninsula, there may be frequent hot winds from south of the equator. However, local land and sea breezes and mountain effects will confuse the prevailing wind pattern.

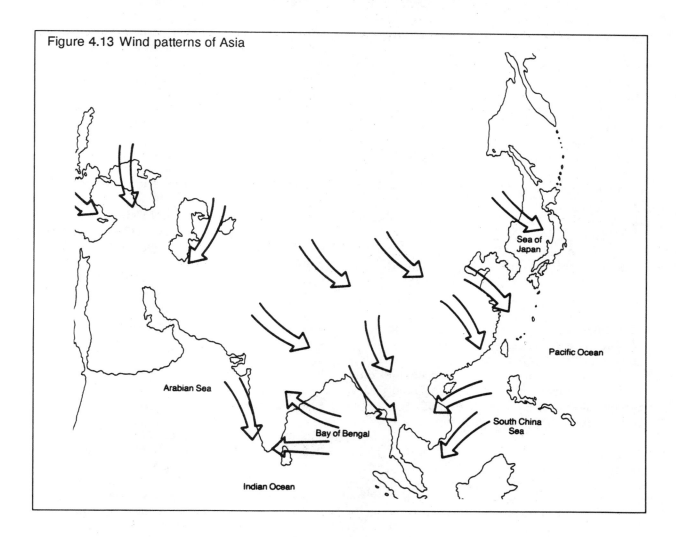

Figure 4.13 Wind patterns of Asia

4.4 Resource Assessment

Information sources

Before investing in a windpump the potential buyer must have some predefined idea of whether the site will be suitable. One of the principal factors in deciding this is the available wind resource. There are three sources of data that can be used for wind resource assessment:

● Meteorological records
● Site measurements
● Local knowledge

All these three methods have their uses, but also their limitations.

Meteorological records may stretch back many years. They can therefore give a broad view of how the wind can change from year to year, and also give a realistic mean wind for each month. More detailed data may also give distributions in terms of direction or of speed. However, it is rare that there will exist a recording station in the vicinity of the proposed site. Even if such data does exist, account must be taken of the local factors that may affect both the meteorological station and the windpump site. Meteorological stations are often located at airports, which may not be representative of the general wind regime. The airfield is likely to be located in a more sheltered area, and the recording mast may be surrounded by grass.

Places to check for meteorological records are:

● Government weather bureau or meteorological office
● Universities doing weather-related research
● Airports and seaports

The only way properly to verify the wind regime for a site is to make measurements at that site. To do this at all thoroughly requires a great deal of time. Even a full year of data gives no guarantee that the year recorded is typical. Fortunately this degree of accuracy is not normally necessary with windpumps.

Manually-recorded data may suffer from being taken at long time intervals, and so may not be representative. It may therefore lack the detail to determine anything of the relative frequency of different speeds or directions of winds. An electronic data logger (Figure 4.14) will provide all the detail needed and avoid the need for human intervention. However, this will involve more costly electronic equipment and will probably need computer facilities to process the data.

If there are other windpump users in your area, the best approach is to ask them about their experiences and the way the windpump performs throughout the year.

Local knowledge from long-term residents of the area is also useful to complement other data sources. They are likely to have an intuitive feel for the way the wind varies with the conditions and the seasons. This can be of particularly use in assessing the effect of the local topography. However, the human mind is notoriously subjective, and using local opinion alone can be very misleading.

A more quantitative approach is to use a limited set of on-site measurements to determine the relationship with the nearest meteorological station. This may involve taking measurements for different wind directions and speeds and comparing them with data from the meteorological station for that time. If some idea is gained of how their windspeeds relate for different wind directions, then the meteorological records can be used to imply what the conditions will be like at the site. It should be stressed that even this method is very rough and prone to error.

If records are available note should be taken of the frequency of occurrence of different numbers of consecutive calm days. This is useful in deciding on storage capacity. Also, if possible, determine the maximum windspeed and bear in mind that the maximum gust may be a factor of 1.5 to 2 greater. You should specify this to the windpump supplier to make sure that the pump will not suffer damage in extreme weather conditions.

Measurements

The windspeed is measured with a device called an anemometer (figure 4.15). These usually take the form of small rotating cups on a vertical spindle that drives either a clockwork totalizer or electrical recording device. The output is usually read as a speed. Some devices are used over a period of time and record the distance of wind that has travelled past. This is called a 'run-of-wind' recorder, and the mean speed is found by dividing the distance by the recording time.

Windspeeds are usually measured in metres per second (m/s), but if data is being used from several sources there are some other units that may be found. The most common units are shown in Table 4.1 together with the conversion into m/s. Windspeeds at ports may be measured on the Beaufort scale, which is defined in Table 4.2. Wind direction is measured with a wind vane.

Anemometers may be either hand-held, or mounted on a mast. They may also need to be read manually or may be electrically connected into an automatic data recorder. When using a hand-held instrument, be sure to hold it still and level above your head at arms length away from your body, side-on to the airflow. Stay clear of obstructions and do not block the wind yourself.

The standard height at which windspeeds and directions are measured at meteorological stations is 10m in most parts of the world,

Table 4.1	Commonly used units of windspeed	
Unit	Abbrev.	x factor to get m/s
Meters per second	m/s	
Miles per hour	m.p.h	0.45
Kilometres per hour	km/h	0.28
Knots	kn or kts	0.51

and it is recommended that site measurements are taken at this height if they are to be compared with meteorological records. It is also wise to check the anemometer height for any meteorological records used, as some stations may monitor at unusual heights. Remember that windspeed increases strongly with height over the first few tens of metres, and so measurements at different heights cannot be easily compared. However, if the 10m wind is known, the nomogram in Figure 4.5 can be used to find the windspeed at any other height for different terrains. It may also be desirable to take readings at the proposed windpump rotor hub height if this is possible.

Table 4.2 The Beaufort Scale

Force	Name	Definition	Speed range (m/s)
0	Calm	Smoke rises vertically	0
1	Light air	Direction shown by smoke, but not vane	1
2	Light breeze	Wind felt on face, leaves rustle	2 - 3
3	Gentle breeze	Leaves, small twigs & flags in motion	4 - 5
4	Moderate	Raises dust & paper, small branches move	6 - 8
5	Fresh breeze	Small trees in leaf sway	9 - 11
6	Strong breeze	Large branches move, wires whistle	11 - 14
7	Near gale	Whole trees move, walking slightly impeded	14 - 17
8	Gale	Breaks twigs off trees, walking difficult	17 - 20
9	Severe gale	Slight structural damage	21 - 24
10	Storm	Trees uprooted, considerable damage	> 25

Using the Information

Once windspeed data has been measured or obtained from records, there are different ways in which it can be used. This will depend on the form of your data, how detailed it is and how reliable and appropriate you think it is. For instance, if the nearest meteorological station were 100km away in an area of different topography, then only a very rough use of its data would be appropriate.

The simplest approach is to try and form a mean windspeed for each month. To do this it is important to have readings taken at various times of the day and night to get a representative mean speed. The greater the number of readings throughout the month, the more meaningful the result will be. If this figure is for the 10m windspeed than this should be converted to the windspeed at the proposed hub height using the nomogram in Figure 4.5

It is sometimes useful to get data on wind direction if possible. This might be needed to decide where the best site is, relative to any obstacles. It will also be useful if topography or the surface type may modify the windspeed in one direction differently to that in another.

Considerations

In summary, a list can be made of the basic steps when assessing the wind potential and deciding on detailed siting:

- Study whatever meteorological records are available that are most applicable to your site. Consult the meteorological office, local recording stations, airfield, seaports and universities. Obtain data on mean speeds, directions, peak speeds and number of calm days.

- Look very carefully at the possible sites, taking into consideration the surrounding topography, positions of buildings, trees and other obstructions. Also note the type of surface cover in different directions, i.e. grass, shrubs, trees etc.

- Take windspeed (and if possible, direction) measurements for the site, and compare these to any local meteorological records. Data over at least a three month period is needed, and if possible all year round. However crude your measurements, they will be of more value than none at all.

- Bear in mind that most windpumps will not start below a windspeed of about 3 m/s and will furl at about 12 to 15 m/s.

- Before final procurement seek the advice on siting of the manufacturer or supplier of the windpump. Otherwise you will be liable and have no right to complain if the windpump performs poorly due to incorrect siting.

Figure 4.14 Typical data logger

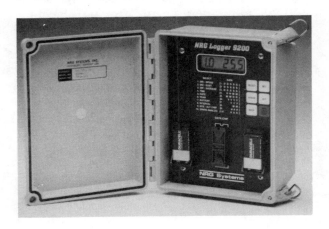

Figure 4.15 Anemometer

Windpump Designs 5

5.1 Principles of Conversion

Extracting energy from the wind

Wind is simply air in motion, invisible but capable of carrying and dissipating large volumes of energy, as is apparent from the damage that is common in the aftermath of any severe storm. Air at sea level has a mass in the region of 1.2 kg per cubic metre, and it is the movement of the invisible air mass, caused by uneven heating of the atmosphere by the sun, that is the source of wind energy.

The nature and characteristics of the wind resource has already been explained in Chapter 4, but the main consideration affecting windpump design (and any type of windmill design) is the extreme variability of the power in the wind in relation to the speed of the wind.

The power in the wind at low wind speeds is too diffuse to be practically exploited (for example it is only about 10 W/m² at 2.5 m/s) yet, by contrast, the power in winds of above 10-15m/s becomes so high that it demands excessive levels of structural strength from a machine (for example 15m/s represents about 2200 watts/m²).

Therefore, most practical windpumps are designed to function in windspeeds in a range from around 5m/s to 10m/s, where the wind blows for a reasonable proportion of time and also reasonable power levels exist.

All windturbines, including windpumps, require a rotor which interacts with the passing airstream and in the process removes some of the kinetic energy from the wind. Therefore there is always a reduction in wind velocity as it passes through a windmill rotor. Obviously if one attempted to extract 100% of the energy from the wind, it would imply stopping the airflow completely which is clearly impossible if energy is required.

There is an optimum level of energy extraction which maximises the energy removed from the wind per unit area. Text books on wind energy prove that the 'perfect windmill' can extract a maximum of 59.3% of the kinetic energy in the wind. In such a situation the wind downstream of the rotor is reduced in velocity to 1/3 of the original free-stream velocity.

In the real world this ideal efficiency cannot be achieved. The better windpumps manage a rotor efficiency in the 30 to 40% range and the best large wind turbines for generating electricity can achieve rotor efficiencies approaching 45%.

Lift and drag

There are two different physical principles by which windmills can remove energy from the wind. These are known by engineers as 'lift' and 'drag' respectively.

Lift is a force caused by the interaction of an area of flat or, more generally slightly curved, material set at a small angle to the airflow. The effect of lift is to generate a strong force perpendicular to the direction of the air flow.

Drag on the other hand is generated in a cruder manner and occurs by obstructing the air flow. Drag forces always act in the direction of the flow and not perpendicular to it.

The most efficient way to extract energy from the wind is to use lift forces rather than drag forces, because simply obstructing the wind causes turbulence in the wake of the blade which represents a useless dissipation of energy.

Figure 5.1 illustrates a cross section through what is known as an aerofoil. The most obvious example of an aerofoil is the wing of an aircraft.

The purpose of its shape is to deflect the airflow in such a way that the flow passing over the top of the aerofoil smoothly follows a longer path than the flow that passes beneath it.

The effect of this is to cause the flow over the top to speed up slightly relative to the flow underneath. As a result of a slight difference in velocity between the air on either side of an aerofoil, there is a higher pressure on one side compared with the other.

Pressure is in fact a force distributed over an area, so the aerofoil consequently experiences a force resulting from the average pressure difference between its two sides. This force is the phenomenon known as 'lift'.

When a typical aerofoil is gradually moved from being edge-on to the wind (i.e. zero angle of attack) to greater and greater angles of attack the lift force steadily increases and the drag force remains quite small until at some angle (around 15° with a well-designed aerofoil) the drag force starts to increase rapidly.

At even greater angles of attack the lift no longer continues to increase and may sharply decline if the angle of attack is increased further. The phenomenon when lift suddenly declines and drag increases is known as 'stalling'.

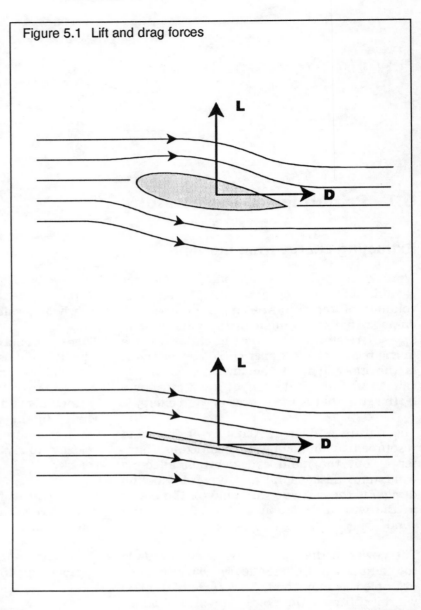

Figure 5.1 Lift and drag forces

A flat plate stalls much sooner than a curved aerofoil, because the sharp leading edge provokes separation of the air flow from the surface of the blade at a smaller angle of attack than if the aerofoil is rounded to direct the flow smoothly along its surface.

In practice a windmill uses a curved or 'cambered' plate which can achieve almost as high a lift force as a solid aerofoil. They are also less costly, as they can be formed from a rolled piece of thin metal sheet. Cambered plates have inherently more drag than a solid aerofoil and are only suitable for slow-running windmill rotors where the drag is not a serious problem. They also have little resistance to bending and therefore need a frame of some kind to support them.

5.2 Rotor Design

Vertical axis versus horizontal axis

Windmills have rotors that most commonly rotate about a horizontal axis (i.e. the rotor shaft is horizontal, or nearly horizontal) but there are designs that rotate about a vertical axis.

Vertical-axis windmills have the advantage that they do not need to be oriented to face the wind, since they present the same cross-section to the wind from any direction. However this is also a serious disadvantage as under storm conditions a vertical-axis rotor cannot be turned away from the wind to reduce the wind loadings on it, as is normal practice with most windpumps. Therefore the rotor of a vertical-axis machine must be able to withstand the most severe windspeeds that nature can throw at it.

There are three main types of vertical axis windmill:

Panamone differential drag devices, (see Figure 5.2). Here the rotor is generally shielded from the wind on one side and fully exposed on the other, so that the exposed part is dragged by the wind and provides a turning force relative to the unexposed side. Such devices were used traditionally in Afghanistan for milling grain, and more elaborate forms were built in China.

Such devices are simple but highly inefficient, so they produce little power in relation to their size. Typical efficiency is around 10%.

Savonius rotor or 'S' rotor devices (Figure 5.3) consists of two or three curved interlocking plates grouped around a central shaft. It works by a mixture of differential drag and lift.

The Savonius rotor has been promoted as a device that can be readily improvised on a self-build basis, but its

Figure 5.2 Panamone drag devices

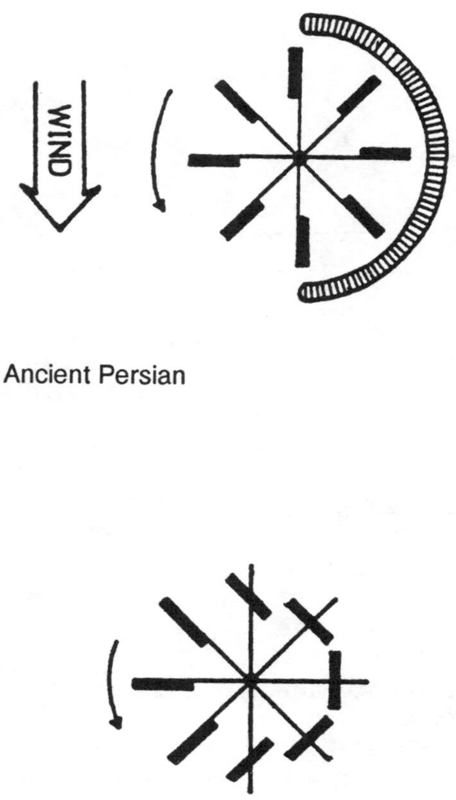

(a) Ancient Persian

WIND

(b) Traditional chinese

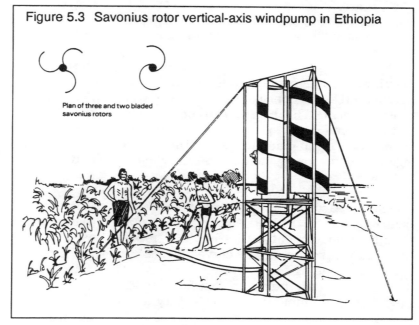

Figure 5.3 Savonius rotor vertical-axis windpump in Ethiopia

Plan of three and two bladed savonius rotors

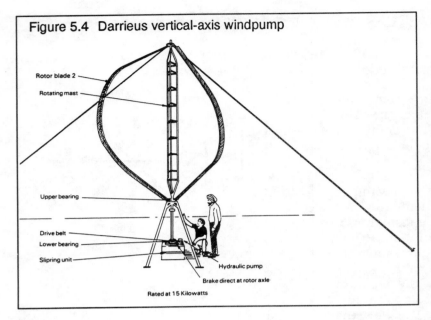

Figure 5.4 Darrieus vertical-axis windpump

Rotor blade 2
Rotating mast
Upper bearing
Drive belt
Lower bearing
Slipring unit
Hydraulic pump
Brake direct at rotor axle
Rated at 15 Kilowatts

apparent simplicity is more perceived than real as there are serious problems in mounting the inevitably heavy rotor securely and in coupling its vertical drive shaft to a positive displacement pump.

The main disadvantages of the savonius rotor are two-fold:

● It is inefficient, (around 15% at best), and involves a lot of construction material relative to its size, so it is less cost-effective as a rotor than most other types;

● It is difficult to protect it from over-speeding in a storm and hence flying to pieces (except with very small examples).

Darrieus wind turbines (Figure 5.4) have aerofoil cross-section blades and are thus quite efficient. The blades can be straight, giving the machine an 'H'-shaped profile, but in practice most machines have the curved 'egg-beater' or troposkien profile as illustrated.

The main reason for this shape is because the centrifugal force caused by rotation would tend to bend straight blades but the troposkien shape taken up by the curved blades can resist the bending forces effectively. Vertical-axis windmills are rarely applied for practical purposes on windpumps although they are a popular subject for research.

Rotor theory

It is outside the scope of this handbook to cover detailed rotor design. Specialized books on the subject should be consulted if such information is required. Also, this section relates only to horizontal-axis rotors as used for the vast majority of windpumps.

Most horizontal-axis rotors work by lift forces generated when 'propeller' or airscrew-like blades are set at such an angle that at their optimum speed of rotation they make a small angle with the wind and generate lift forces in a tangential direction. Because the rotor tips travel faster than the roots, they 'feel' the oncoming wind at a shallower angle and therefore an efficient horizontal-axis rotor requires the blades to be twisted along their length so that the angle with which they meet the wind is constant from root to tip.

As a general principle there is a relationship between the speed of a windmill (relative to the wind) and the proportion of the area swept by the rotor that needs to be 'solid'.

The ratio of the speed of the tips of a rotor to the wind is known as the 'tip-speed ratio' (usually written as the Greek letter lambda, λ). The 'solidity' of the rotor can be defined as the ratio of the total blade width or chord to the circumference of the rotor.

For example, if the tip speed ratio $\lambda = 3$, then the tips of the rotor blades travel at three times the speed of the wind. For optimum efficiency with a given solidity there is an optimum tip-speed ratio as indicated by the relationship in Figure 5.5.

It is noticeable that windpumps tend to have high-solidity (and hence low-speed) rotors while wind-electricity generators tend to have low-solidity and high-speed rotors. The main reasons for this are that wind-pumping generally requires a pump to be driven at a low pumping rate while a generator needs to be driven at high speed.

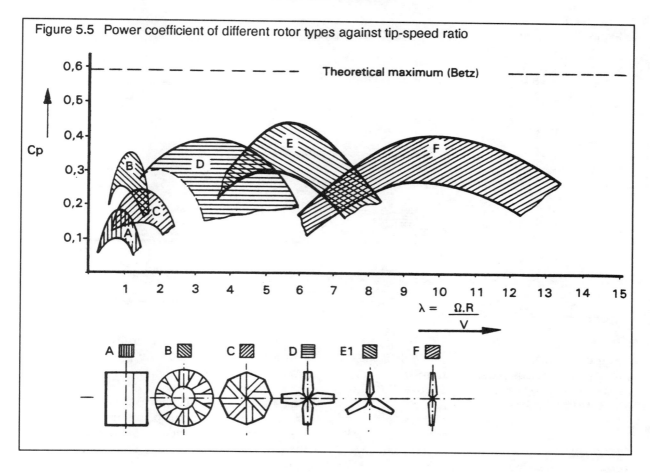

Figure 5.5 Power coefficient of different rotor types against tip-speed ratio

Power and torque characteristics

A windpump, requires a certain lowest windspeed (V_{start}) in order to initiate rotation. At that windspeed the turning force or 'torque' of the rotor is just sufficient to exceed the resistance to movement imposed by the pump.

As the windspeed increases, the power available increases steadily until the nominal rated wind-speed (V_r) is reached, after which point measures are taken to limit any further extraction of power from the wind.

To prevent damage at high windspeeds the windmill is usually 'furled' (shut down) at V_f. When the windspeed

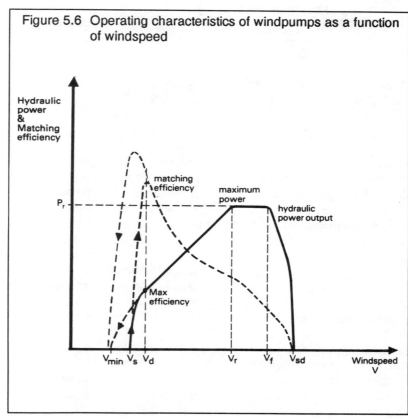

Figure 5.6 Operating characteristics of windpumps as a function of windspeed

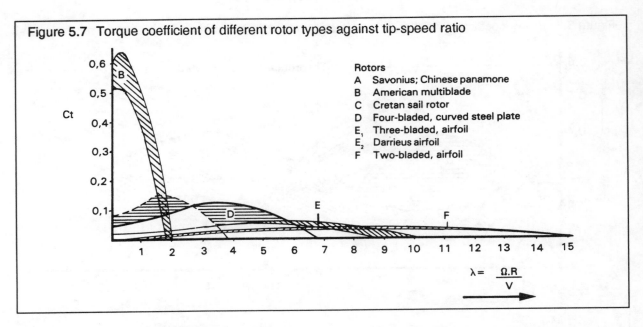

Figure 5.7 Torque coefficient of different rotor types against tip-speed ratio

Rotors
A   Savonius; Chinese panamone
B   American multiblade
C   Cretan sail rotor
D   Four-bladed, curved steel plate
E_1 Three-bladed, airfoil
E_2 Darrieus airfoil
F   Two-bladed, airfoil

$$\lambda = \frac{\Omega.R}{V}$$

falls the windpump can continue to operate at a minimum windspeed (V_{min}) which is below the starting windspeed, because the starting inertia has been overcome. Figure 5.6 illustrates this relationship.

The efficiency of a windmill rotor is generally described by a non-dimensional parameter known as the Coefficient of Performance (or Power), C_p. This can be defined mathematically as follows:

$$C_p = P/0.5\rho AV^3$$

The C_p is the ratio of the shaft power, P, to the instantaneous power in the wind of velocity V through cross-sectional rotor area A. Figure 5.7 indicates some typical performance curves for various types of windmill rotor in which C_p is plotted against the tip-speed ratio (lambda, as previously described). Any type of rotor has a characteristic curve of this kind, with the optimum efficiency achieved at a given, fixed tip-speed ratio.

So far as windpumps are concerned a critical factor is 'starting torque'. This is the torque the rotor needs to initiate pumping and it determines the windspeed needed for a windpump to start. It can be seen from Figure 5.7 that a high-solidity rotor tends to have a high starting torque coefficient which increases slightly as the windmill starts to speed up and then declines sharply. A low-solidity rotor, on the other hand, has a low starting torque

coefficient and the torque builds up to a peak at around two-thirds of the tip-speed ratio for optimum efficiency.

The implication of this is that a typical conventional windpump rotor can have a starting torque coefficient (and hence actual starting torque capability) as much as 20 or 30 times greater than a low-solidity propeller type of rotor as found on a typical wind electricity generator. For driving typical piston pumps this is a key factor and more important than the slightly reduced maximum efficiency applicable to high solidity rotors.

5.3 Windpump Performance Theory

Consideration of a windmill rotor in isolation provides little indication of a windpump's performance. In the end it is the interaction between the rotor and the pump, via the transmission mechanism, that dictates the performance of the system.

For a piston pump to start pumping, the windpump crank needs to exert a force on the pump rod sufficient to lift the weight of the pump rods and the piston, plus the weight of the water sitting on the closed valve in the piston, plus overcoming friction. The amount of water to be lifted is a column of water equal in diameter to the piston and of a height equal to the head, which is the depth from the point of delivery to the surface of the water in the well. The throw of the crank (or eccentricity) is equal to half the stroke of the pump.

The rest position for the windpump under perfectly calm conditions will normally be with the piston at bottom dead centre. As the wind increases it will try to rotate the rotor and hence lift the crank. When this happens the load will increase until it reaches a maximum when the crank is at right-angles to the direction of load from the pump rods, i.e. the crank is horizontal. If there is sufficient starting torque generated then the crank passes the horizontal position of peak torque and the load decreases as the effective eccentricity decreases sinusoidally towards top-dead centre.

After this point the weight of pump rods plus water acts to pull the rotor round and the valve opens in the pump as the piston descends to admit more water. There is zero load (or a small negative load) on the pump rod as the crank moves ready to start a new cycle.

Once the rotor succeeds in lifting the crank past the point of maximum resistance it tends to accelerate because the load actually decreases and builds up momentum. Therefore once the rotor is rotating, momentum smooths the load and less wind is needed to maintain operation than was needed to initiate it in the first place. In fact, the mean torque requirement is approximately 1/3 of the peak·torque requirement and as torque is proportional to the square of the windspeed, the minimum wind to maintain operation is only about 0.6 of the starting windspeed.

The torque requirement of a piston pump is constant regardless of the speed of operation.

Figure 5.8 shows (a) power and (b) torque as a function of windspeed for a typical multiblade windpump rotor, with the pump load curve superimposed. The system will follow the pump load curve (which for piston pumps is a straight line in both cases).

The torque graphs show how a windspeed of, say, 3m/s may be needed to start the windpump. However, they also show that for the reasons given above, once rotation has commenced the windspeed can fall (theoretically) to as low as 1.8m/s (i.e. to 0.6 of 3.0m/s) before the windpump will stop.

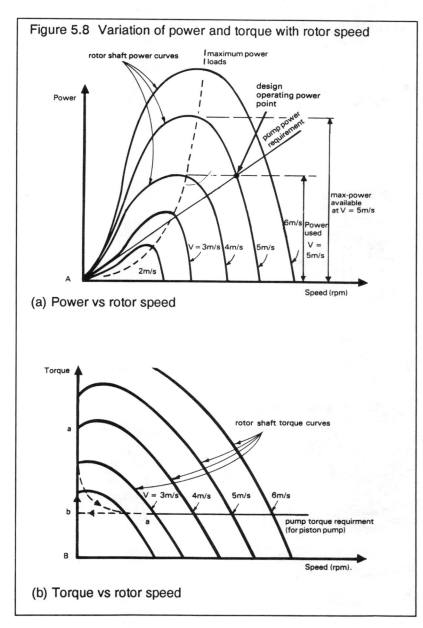

Figure 5.8 Variation of power and torque with rotor speed

(a) Power vs rotor speed

(b) Torque vs rotor speed

The size of the pump effectively dictates the starting windspeed for a given windpump and pumping head.

This is because the bigger the pump, the larger the starting torque requirement and hence the higher the windspeed needed to start the rotor. Once started however, bigger diameter pumps have greater output because more water is delivered per stroke.

Generally speaking, windpumps are fitted with pumps that will allow them to start in a windspeed of approximately three-quarters of the local mean windspeed. This has been found to be a rule of thumb that results in near to the best possible compromise between getting the windpump to run frequently enough and achieving a good output in stronger winds.

In any case most manufacturers will provide recommendations on pump sizing and it is generally best to adhere to these.

5.4 Transmission Options

Conventional transmissions

The key part of a windpumping system in terms of ensuring its general robustness and reliability is the power transmission train which carries the forces generated by the rotor through to the pump.

A windpump subjects its mechanical components to an unusually large number of load cycles during its lifetime. A typical windpump running on average for 15 hours per day for 25 years will run approximately 136,000 operating hours and complete in the region of 160 million pump strokes.

When it is considered that a high quality car only completes some 3000 to 5000 operating hours before it is scrapped it indicates that the mechanical requirements for a windpump are far from trivial.

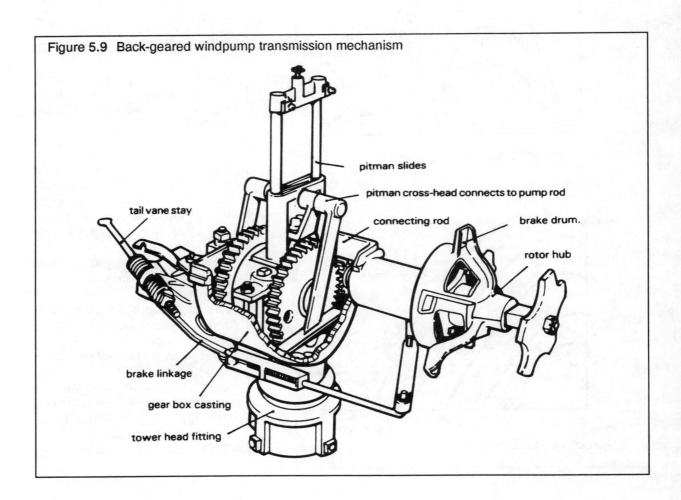

Figure 5.9 Back-geared windpump transmission mechanism

pitman slides

pitman cross-head connects to pump rod

tail vane stay

connecting rod

brake drum.

rotor hub

brake linkage

gear box casting

tower head fitting

This is one reason why 'home made' windmills so often fail, because they lack the essential safety margins and good engineering design to survive the continuous pounding of components that they need to endure during their lifetime.

Geared vs direct drive

In most cases, windpumps are designed to drive a piston pump at quite low speeds, no greater than 40 or 50 strokes per minute. A high-solidity rotor of the type appropriate for running a piston pump has a tip-speed ratio between 1.0 and 1.5 in most cases (see Section 5.1) and a simple calculation can be done to show that given a furling wind speed typically of 12m/s there is a minimum rotor size that will run slowly enough to drive a pump directly without exceeding the 40 to 50 strokes/minute maximum pump speed.

In general rotors smaller than around 4.5m in diameter tend to run too fast for 'comfortable' operation.

The mechanical drive train in most smaller windpumps of the conventional type is therefore 'back-geared', i.e. the main shaft drives a small gear which meshes with a larger gear (see Figure 5.9). Gear ratios of 0.5 to 0.25 are most common. The second larger gear turns more slowly than the rotor, and has a crank-pin which generally links with a slider or Pitman that runs up and down two vertical guide rails mounted over the gearwheels.

The slider in turn is linked by a vertical rod projecting downwards to the pump. Generally the two shafts, the primary drive shaft and secondary low-speed shaft, run in plain bronze bearings and the entire mechanism is enclosed in a large cast iron container partially filled with lubricating oil.

Sometimes a small oil pump is built in to the system to force oil up the connecting rod to the pitman slider and sometimes a splashing arrangement is used to spread the oil over all moving parts.

When reduction gearing is provided, the rotor can happily run at speeds up to 200 rpm or so without the pump speed ever reaching more than 50 strokes per minute and this is

probably one of the reasons for the high reliability and long life commonly achieved from this kind of machine.

The main reasons for limiting the pumping speed to 50 strokes per minute are as follows:

- Cavitation and shock loads in the water supply can cause damage and hammering in the system if it is attempted to accelerate the column of water in the rising main too rapidly.

- At speeds greater than about 60 strokes per minute the rotor and mechanism can attempt to accelerate the pump rod on the down stroke instead of letting it simply fall back under its own weight. This can actually buckle the pump rod by pushing it downwards faster than it would naturally fall.

- Even if the pump rod does not buckle through being driven downwards, reversing the loads on all the bearings, links and connections causes any slack components to hammer each other and hence to wear far more quickly than if they are only loaded from one direction.

Large diameter windpumps (where the rotor generally is limited to speeds of less than 50 rpm on account of its size) are commonly made with a direct-drive mechanism (Figure 5.10). A few smaller machines with surface pumps or shallow well pumps where the pump rods are too short to buckle easily also use direct-drive mechanisms.

Sometimes the crank connects directly with the pump rod, but this is only possible in situations with a small crank throw and hence a short pump stroke. With large machines it is normal to have an overhead radius arm or slider so that there is no significant sideways motion of the pump rod where it emerges below the mechanism and passes through the narrow space in the yaw tube which engages the mechanism with the tower head.

A novel type of transmission mechanism is used in the new I T Power small windpump that allows better pump to rotor matching and is thus very efficient in light winds (Figure 5.11).

Figure 5.10 Direct drive mechanism (*Kijito* windpump, Kenya)

Many of the smaller and more modern machines use grease-lubricated ball bearings rather than oil-lubricated plain bearings and this leads to a much lighter but possibly less robust mechanism.

Smaller direct-drive machines, although often quite inexpensive and simple, will normally not achieve anything like the long life and reliability of the more traditional designs.

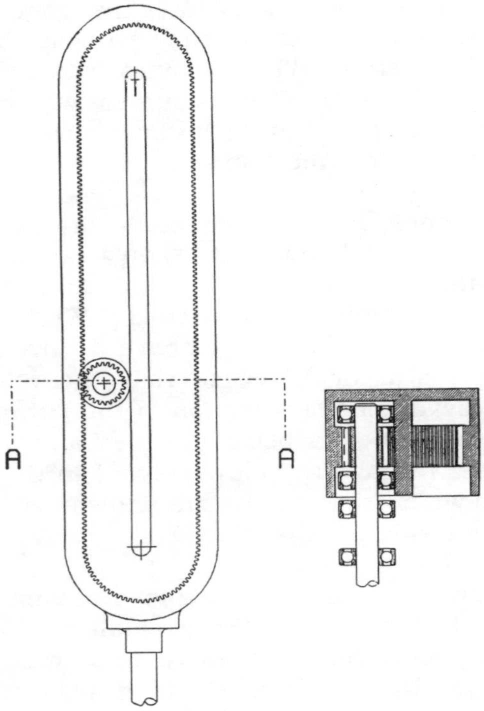

The pump rods (sometimes known as sucker rods) couple the windpump mechanism to the pump. It is possible to use standard pipe, solid steel rods or wooden rods. Wooden pump rods have the advantage that they float in the rising main and therefore impose less of a load on the windpump at start-up (the rotor is not required to lift their weight as it is with non-floating pump rods). However they are generally more expensive than steel pump rods since special quality close-grained hardwood timber, which is completely knot-free and which retains its strength when permanently saturated has to be used. Home-made wooden pump rods made from dubious quality timber are not a good idea.

When using wooden pump rods it is usual to use wrought iron couplings, often fitted with quick-release joints which greatly speed up the process of raising and lowering the pump piston for servicing compared with screwed fittings.

5.5 Remote Pumping Options

Virtually all windpumps of the conventional type have to be located directly over the borehole or well so that their pump rods hang exactly vertically and connect directly with the borehole rising main. Although this works well, it has two major disadvantages:

- wells and boreholes (and especially surface water sources) tend to be located on low ground, often with trees nearby. Hence the best locations for accessing ground or surface water tend to be the poor locations for accessing the wind.

- the structure of the windmill and its tower can make it difficult and inconvenient for accessing the borehole to overhaul the pump or to carry out repairs.

Figure 5.11 I T Power small windpump transmission

Therefore, conventional windpumps have tended to be limited to relatively flat and arid regions and many opportunities for using them in other places have been lost due to the relative inflexibility that dictates their siting. One way around this problem is to use a windmill capable of being located at some distance from the water source. Broadly speaking there are four types of transmission medium that may be used to transmit power from a windmill to a remotely located pump:

- Electrical
- Pneumatic
- Hydraulic
- Mechanical

Electrical transmission

A wide selection of wind-powered electricity generators are commercially available and electric pumps can readily be procured. However, the interfacing of wind generators with electric pumps is not straightforward and should not be attempted by the user as some companies sell wind-electric pumping systems as standard products (Figure 5.12).

There are several combinations of wind electric system that are possible. An AC wind generator can power either a DC pump via a rectifier, battery and regulator or can power an AC pump via a rectifier, battery and inverter. The batteries serve two purposes: to smooth the variation of output from the wind turbine and to give some storage capacity. Special deep-discharge batteries need to be used (vehicle batteries are not suitable).

An AC induction generator can also be used to synchronize with and hence drive an induction motor and pump directly. The motor will run at similar speeds to the generator so it needs to be coupled to a centrifugal pump with a characteristic capable of handling a range of speeds with reasonable efficiency at the required head. The better systems use permanent mag-

net generators (PMG) which combine a very high efficiency (up to 95%) with a low speed of rotoation, and hence are suitable for direct coupling. These systems may be useful for future large applications, as they are well suitedfor low-lift and high-volume pumping where mechanical windpumps may be limited by pump diameter and their inability to be scaled-up above a 10m rotor size.

In general there are a number of difficulties associated with wind-electric systems:

- Only a few manufacturers offer an integrated 'packaged' wind electric pumping system (see the Buyer's Guide section) and there is only limited experience with the technology.

- Typically a wind-electric pumping system will be half as efficient as a good mechanical windpumping system. The ability to locate the windmill at a windier location may compensate for this to some extent.

- If low transmission voltages are used (12-48V) then thick and expensive cables are needed to prevent resistive losses. This can be avoided using by higher voltages (220-380V), but then professional standards of electrical engineering are needed to avoid safety hazards, which also adds expense.

Figure 5.12 Wind electric pumping system

In conclusion, other than the simple option of running a low-voltage electric pump from batteries charged by a small wind generator, it is not recommended that wind-electric systems be used for pumping except where they are properly packaged and integrated by professional suppliers.

Pneumatic transmission

Pneumatic transmission is relatively inefficient as the process of air compression, transmission down a hose-pipe and then expansion through a slave pump actuator, pneumatic pump or an air-motor involves a number of inefficiencies.

The advantages of low pressure compressed air transmission are that the transmission line can be low-pressure plastic pipe or hose, which is less expensive than cable. Also the transmission is 'soft' so that the pump is not mechanically linked to the windmill. This means the pump can operate slowly and the windmill can run fast without any shock loads.

One range of pneumatic windpumps has been commercially marketed in recent years, the *Bowjon*, made in the USA. This is in fact a light-weight windpump coupled to a small industrial air-compressor which can be used either to drive an airlift pump or an industrial pneumatic pump.

Quite a lot of promising experience has also been gained with pneumatic windpumps in Brazil where the characteristics of air-lift pumps happen to suit the well characteristics. Air-lift pumps are extremely simple and virtually maintenance free.

Air-lift pumps rely on the principle that if compressed air is introduced into a rising main, it produces a froth of water which rises up the pipe because the density of the mixture of air and water is lower than the density of plain water. However they need a deep borehole in relation to the static head (i.e. a considerable depth below the rest water level).

An alternative to the air-lift pump for a pneumatic system is a pneumatic displacement pump. A pump of this kind requires an air supply of approximately the same pressure as the static head and is best suited to low-lift surface water pumping at heads from around 5 to 20m (approximately 0.5 to 2 bar of air pressure).

For field workers and end-users discussion of pneumatic remote pumping systems must remain somewhat academic as, at the time of writing there is little choice in terms of commercially available systems.

Hydraulic transmission

There are two possible forms of hydraulic systems; a two pipe system and a one-pipe system.

In the two-pipe system the hydraulic fluid, either oil or water, is pumped down one pipe, drives a slave pump actuator and the fluid is then discharged from the slave device and returned up a second pipe to the source.

The one-pipe system involves a reciprocating flow in which the fluid acts as an incompressible 'push rod' by pushing a remote actuator to pump some water, which is then returned by a spring or mass.

A hydraulic transmission using a *Kijito* windpump has been successfully demonstrated in Kenya. It should be noted that there are also at least three manufacturers of handpumps which use a one-pipe water hydraulic transmission system, namely Vergnet, Omega and Merrill. Such pumps could effectively be used with a small windpump, providing the windpump was not so fast-running as to cause water hammer or other dynamic problems.

Mechanical transmission

Numerous attempts have been made to mechanically operate a pump located at a distance from the windpump, but the mechanisms are clumsy and expensive because of the large forces involved. They also tend to be unreliable.

Mechanical transmissions for remote pumping over any significant distances are not commercially available and thus not considered here.

5.6 Windpump Components

The key components of the conventional geared windpump are described here in order for the reader to gain a better understanding of this widely used windpump type.

Head and rotor

The rotor is invariably fabricated from hot-dip galvanized mild steel components. The blades themselves tend to be rolled from thin steel plate which is subsequently galvanised.

Figure 5.13 illustrates a typical windpump rotor assembly. Two concentric rings of flat bar held on brackets are attached to 'A'-shaped radial spokes. The cambered plate blades are bolted to brackets attached to the concentric rotor rings. An important requirement is that all the rotor nuts and bolts are securely tightened on assembly, as any component coming loose within the rotor can cause catastrophic results if, for example, it catches against the tower.

Generally a high solidity is used so as to achieve good starting torque combined with slow rotational speeds.

The smaller windpumps have reduction gearing for reasons previously explained. Reduction gear machines have twin shafts. The main shaft which carries the rotor has a small pinion which usually drives a pair of gearwheels carried on a secondary low-speed shaft which drives a crank/connecting rod system linked to the pump rod.

The gearwheels are often paired so that the heavily loaded crankpin which drives the pump rod is located between the two gears and is thereby more evenly loaded than if it projected from a single gearwheel.

Machines with rotors larger than about 5m in diameter are usually direct drive and these generally have the crank at the rear end of the main shaft and drive the connecting rod, pitman and pump rods directly.

The rotor is normally assembled onto a massive cast or forged hub which in turn is keyed or clamped onto the main shaft. The main shaft is most often carried in plain bearings (i.e. bronze or white-metal lined bushes) housed in a large steel or (more usually) cast iron transmission casing. This casing not only acts as an oil reservoir but it is also the chassis member which holds the key components of the windmill together.

A yaw tube normally extends downwards from it through bearings in the tower top. Brackets at the back of the casing provide attachment points for the tail-vane assembly.

The pitman slider extends above the casing and is generally protected from the elements by a sheet metal box (most commonly formed from galvanized steel sheet). This is usually simply lowered onto the head mechanism like a large lid.

In most cases a small oil pump is driven by a cam from one of the shafts and this forces the oil through the primary and secondary shaft

Figure 5.13 Conventional windpump rotor

bearings and also up to the top of the connecting rod to lubricate the pitman slider and the connecting rod small end.

Only relatively modern designs incorporate ball bearings or roller bearings instead of plain bearings. So called 'rolling element' bearings (i.e. ball and roller bearings) have only become widely used and inexpensive since the 1920s. Similarly grease lubrication is also a recent development as until the 1920s reliable grease lubricants were not generally available.

Tail and furling arrangements

The tail vane is normally attached to lugs at the back of the main transmission casing. The tail generally consists of a steel arm pivoted at the rear of the windpump head with a galvanized steel sheet attached to its far end. An upwardly inclined strut attached to a higher point of the main transmission casing is usually provided to stop the vane from sagging under its own weight. The tail maintains the rotor facing the wind but is also often the best way to identify the manufacturer and model of windpump as it is normally used to carry the trade-name.

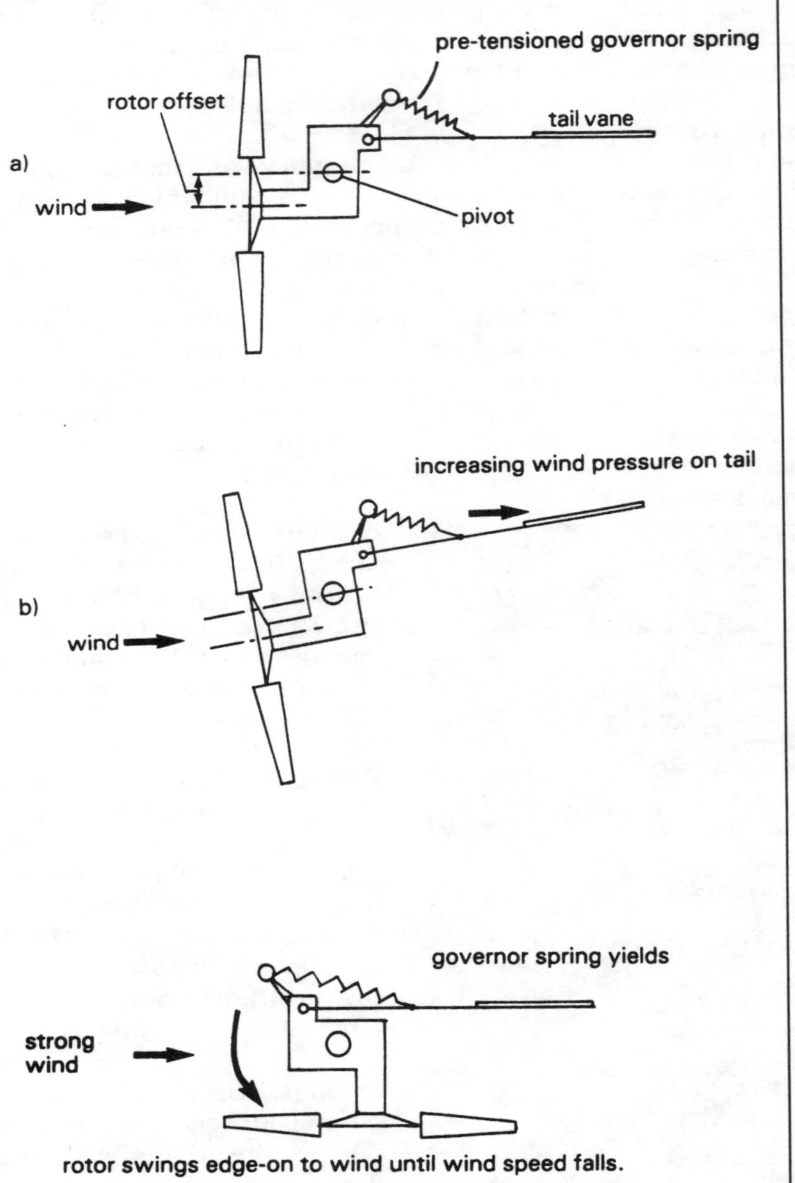

The tail is a key part of the mechanism for providing automatic protection from damage by strong gusts of wind. Figure 5.14 illustrates the principle by which this works.

Note that the rotor shaft is slightly off-set to one side from the yawing axis, in such a way that if the tail vane was not present the rotor would tend to yaw round. However, as soon as the rotor yaws a small amount, the tail vane takes up a small angle to the wind which prevents the rotor yawing any further (as in (a) in the figure).

The tail vane is hinged to the back of the windpump transmission and held in the normal position at right angles to the rotor by a stretched spring holding it against a stop, in much the same way as a spring-loaded, self-closing door.

As the wind increases in velocity it tends to force the windpump to yaw slightly and this increases the wind load on the tail vane considerably (as in (b)). At a certain predetermined windspeed, the load on the tail vane reaches a level

Figure 5.14 Furling principle for automatic yawing

pre-tensioned governor spring

rotor offset

tail vane

a)

wind

pivot

increasing wind pressure on tail

b)

wind

governor spring yields

strong wind

rotor swings edge-on to wind until wind speed falls.

where it can overcome the tension on the spring. When this happens the spring stretches and allows the entire rotor and transmission mechanism to yaw around until the rotor disc lies parallel to the tail vane (as in (c) in the figure).

Usually the design is such that as soon as the spring tension is overcome and the head mechanism starts to move, its leverage increases. In this way the initiation of the furling process is made decisive (otherwise there would be a tendency for the machine to partially furl and unfurl). Some machines have better furling characteristics than others and a potential source of damage can be hesitant or delayed furling resulting from poor design.

An unfortunate feature of this system is that unfurling, which of course takes place automatically as soon as the wind drops to a predetermined safe level, (typically about 5 or 6m/s) can be much more violent than furling. This is especially true if the wind falls suddenly from a high level to a low level then the spring pulls the rotor back round to face the wind and sometimes the tail-vane can impact quite violently against the stop or whatever is provided to prevent it overshooting the unfurled position. The better designs of windpump have a spring buffer of some kind to absorb the impact when the windpump unfurls so as to avoid the potentially damaging crash, which can be much like a large steel door or gate slamming.

Normally the spring tension is adjustable so that the tighter the spring the higher the windspeed needed to cause furling. Most windpumps are set so that they will automatically furl in a windspeed in the range from about 8 to 12m/s. If the spring breaks (and they tend to be prone to rust and eventually break) then the system is 'fail-safe' in that it simply furls at a very low windspeed if no spring tension is present.

In some models of windpump, gravity is used instead of a spring. The most common way to achieve this is to set the pivot for the tail at an angle backwards (and sometimes slightly sideways) from the vertical so the tail tends to hang straight out behind the windpump like a badly hung gate. Then if the rotor is yawed partially by the wind, the tail is forced to lift

slightly as it folds. Therefore an increasing wind load on the rotor is necessary to overcome the weight of the tail. Similarly, if the wind drops, then the rotor is no longer forced around against the tail which allows it to fall back to the straight-behind position.

Manual furling

Apart from the provision of automatic furling in the way described, it is usual to provide some method for manually furling a windpump if it is required to stop the machine. This might be necessary because the water is not needed or because it is planned to do some maintenance or repairs.

The usual way to achieve this is to have a linkage that allows a person at ground level to pull a cable or chain which in turn pulls the rotor around into the furled position parallel with the tail vane. Usually the 'pull-out' simply stretches the furling spring, or lifts the tail in much the same way as a high wind does, but a clever mechanism used on some machines simply reduces the furling spring tension to zero by releasing the spring fixing at the head of the windmill. Then the wind will furl the machine even in a light breeze.

The manual pull out generally involves attaching a chain or cable to the tail vane arm and this is threaded around some pulleys and taken alongside the pump rod through the yaw tube so that it can be accessed from within the tower. There is usually a rotary connection fitted around the yaw-tube which allows the part of the linkage inside the yaw-tube to yaw with the windmill head mechanism as the wind direction changes, but which can be engaged by a fixed component attached to the tower that does not yaw.

The details of how this is achieved vary with different designs, but the fixed component is usually linked by a cable either to a small hand-winch with a ratchet to stop it winding back, or a long wooden lever, attached to the tower near ground level, which can be easily activated to pull-out the windpump and stop it from operating.

Some manufacturers even offer float-operated mechanisms as an optional extra. These can be installed in the storage tank so that when the water level reaches a predetermined level they activate the pull-out and stop the windpump. Then when the water level falls due to usage, the float mechanism drops and releases the pull out so as to re-start the windpump to top up the tank again.

The better quality designs also incorporate a brake, usually a band with friction material on it that can be tightened so as to grip around a drum at the rear of the rotor hub. It is usual for the brake to be linked to the pull-out mechanism so it is only applied once the machine is in the fully furled position and the rotor has stopped.

Any attempt to stop a windpump from running simply by using a brake is likely to achieve no more than burnt-out friction pads so the brake should never be engaged except when the rotor is fully yawed.

Tower

Most farm windpump towers are made from hot-dip galvanized steel components bolted together. The towers generally use angle iron for the verticals and can have three or four legs. Three is marginally more economical in steel usage but tends to be more complicated due to the plan view being triangular rather than square so that special fittings are needed to join the 90° angle-iron flanges to horizontal members joining at 60°. Horizontal members are provided at intervals of around 1 to 2m and are made from the same section angle iron as the legs. Cross bracing is generally essential and needs in some cases to be tensioned.

The tower legs are mainly joined by steel fishplates (like railway tracks) and a large inventory of nuts and bolts is involved with most designs.

Wood has traditionally been used for windpump towers in the USA, and has the advantage of creating a quieter machine (steel towers tend to amplify the clangs and rattles of the mechanism). However, high quality structural timber is necessary and must be treated against rot, insect infestation, etc., so it is doubtful if much cost saving can be achieved in this way.

Figure 5.15 Windpump tower showing access ladder

Some manufacturers sell an adapter tower top to allow their windpump mechanism to be installed on another manufacturer's tower or on a home-made tower.

Many windpumps have a small platform mounted on the tower just below the rotor. This serves two purposes: firstly it provides a surface on which to stand and place tools when undertaking the annual lubrication and overhaul, or other repairs. A second purpose is to provide a guard to prevent a person climbing the tower from accidentally making head contact with the turning rotor, since a conscious effort is needed to look up and climb over the edge of the overhanging platform. Obviously it is in any case preferable to furl or pull-out a windpump, and apply the brake, before climbing the tower, or to wait for calm weather, but occasionally if, for example, the furling mechanism needs urgent repair, this may not be possible.

There is usually a ladder or rungs attached to one of the corner posts to assist anyone needing to climb the machine, although it is common to start the ladder some distance above ground level to deter children or unauthorised people from attempting to climb it.

Piston pumps

The piston pumps used on standard farm windpumps are highly efficient provided that they are used at heads of more than around 10 metres. A piston pump used on heads in the 20-50m range will easily achieve an efficiency better than 75% and at higher heads efficiencies running to over 90% can be achieved. Piston pumps also lend themselves to being driven at low speeds.

A typical borehole piston pump (Figure 5.16) has a drawn brass cylinder liner and a piston with leather cup-seals. The piston body is hollow so that water can pass through it and a disk valve allows water to pass upwards through it on the down stroke, but closes so as to lift the water above it on the upstroke.

An important point to note is that the brass cylinder is fitted inside a steel pipe which in turn screws into a rising main of slightly larger internal diameter than the bore of the pump cylinder. Therefore it is possible to pull the piston up through the rising main in order to change the seals without having to lift the rising main.

The piston is the main wearing component: Typically the seals of a pump with a 300mm (1ft) stroke will slide a total distance of 2000 to 3000km per year (assuming 7 to 10 million pump strokes per year). Not surprisingly even the best seals need replacement at intervals ranging from 1 to 2 years. Certain seals made from synthetic leather are claimed to last up to 5 years under favourable circumstances (i.e. clean and silt-free water).

Pumps with extractable footvalves and pistons are preferable for obvious reasons in deep boreholes, but non-extractable pumps can readily be used in shallower-lift applications.

Piston pumps are less effective at low heads, since the lower the head the higher the flow and the higher the flow the larger the pump diameter. Therefore a rather clumsy large

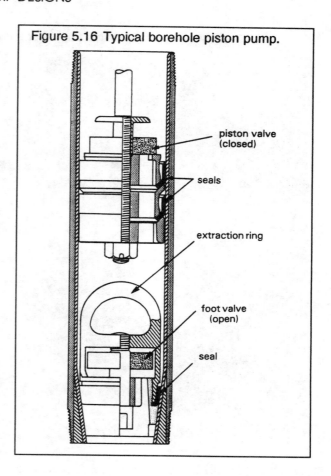

Figure 5.16 Typical borehole piston pump.

piston valve (closed)

seals

extraction ring

foot valve (open)

seal

diameter piston pump becomes necessary at low heads (Figure 5.17) and also tends to be less efficient because passing a high flow through valves results in significant turbulence and head loss.

Other types of pump

Various rotary pumps have been tried with windmills. It is beyond the scope of this book to enter into great detail about pumps, however the main types of pump tried on windmills are the progressive cavity screw pump, or 'Mono pump' (Figure 5.18) which requires a rotating drive shaft.

This type of pump tends to be extremely durable and runs for long periods between maintenance but it is less efficient than a piston pump and significantly more expensive. It also often needs to be geared up from a windmill rotor to run at quite high speed (200-600rpm) and needs high breakaway torque to start it, which is not a good feature with a windmill.

Figure 5.17 Low-head surface piston pump components

Figure 5.18 Progressive cavity pump

Drive shaft

Helical-screw rotor

Flexible stator

Metal casting

Various windmills have been built to drive centrifugal or rotodynamic pumps.

Like the progressive cavity pumps these types also need a high-speed rotary drive and are probably best suited for windmills operating at very low heads and handling large volumes of water.

In general centrifugal and rotodynamic pumps tend to be less efficient than piston pumps by a significant margin.

Traditional Chinese and Thai windpumps (Figure 2.7) have used the so called 'dragon spine' or ladder pump in which an endless belt with wooden plates between the links is dragged up a trough by a suitable mechanism.

These pumps, although seemingly crude, have been in use for centuries and they also achieve surprisingly high levels of efficiency (50 to 70% at heads of only 0.5 to 1m have been recorded). This type of pump can only be used at very low heads as it has to be inclined into the water at a shallow angle, but it is surprisingly effective under the right conditions.

The modern equivalent of the dragon-spine pump is the screw pump which achieves a similar effect by screwing pockets of water up a semicircular trough instead of pulling them up a rectangular trough.

A low-lift windpump using a screw pump to lift water through heads in the range up to 2.3m has been successfully developed in China.

In the last analysis, the majority of windpumps are used at relatively high heads and use conventional piston pumps. Perhaps the reason windpumps are not more widely used for low-head applications is precisely the technical problem of finding a reliable, efficient and easily installed pump for such purposes.

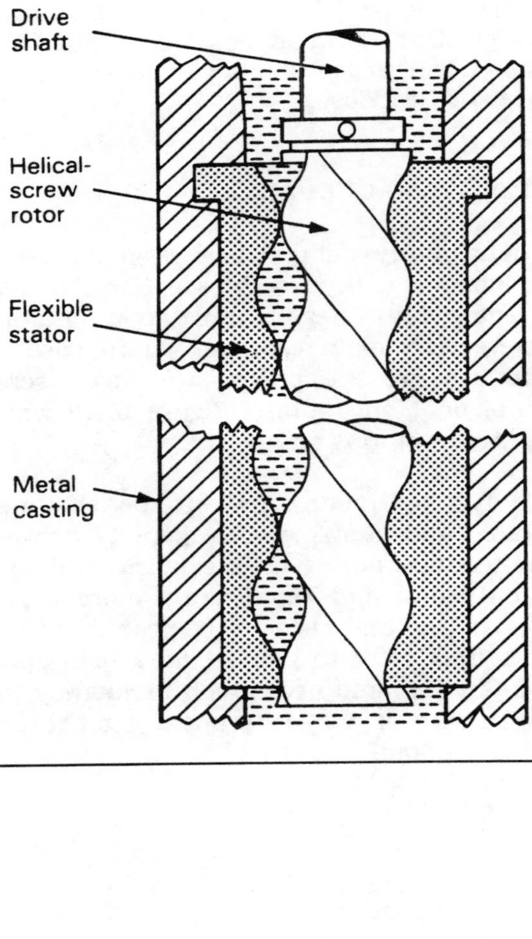

Commercially Available Equipment

6

6.1 Types of Windpumps

Windpumps may be broadly divided into five categories, depending on their design and their method of construction. The class of pump that is most suitable for a given situation will depend on many factors, including the wind regime, the site details, the available budget and the level of reliability that is necessary. In a developing country it will also usually depend on the local level of mechanical capability and which types of windpump are available within the country.

Conventional gear-driven windpump

This is the most common type of windpump (illustrated in Figure 6.1), which has become well established through over a century of use world-wide. Sometimes also called the American Farm Windpump, this type became popular for stock watering in the late nineteenth century in the US. Since then it has spread to almost all countries where windpumps are made, and has been improved and modified by manufacturers for better reliability and performance.

The rotor is multi-bladed, and would typically have between 12 and 30 blades. However, the most important characteristic of the machine is the gearbox. This reduces the speed of the pump-rod action, while increasing the force with which it pumps. This leads to a far more versatile machine, suitable for shallow or deep-well pumping. It also means a more reliable machine, as there is less wear-and-tear with a slow pump action. The gear wheels usually run in a bath of lubricating oil, which must be topped-up by the user every few months. Otherwise, maintenance is minimal.

To manufacture a gearbox requires the facilities to cast metal and cut gears, which adds complexity to the manufacture of gear-driven windpump compared to those with a simpler transmission system. In addition, small volume production is less likely to be economic, as large batches of castings need to be produced.

The windpump is constructed entirely of metal (including the tower) and can only be produced in a well-equipped workshop by machinists with some degree of skill. This means that in general the quality of components and construction is high, as is its reliability. A well-made gear-driven windpump should have a lifetime exceeding 20 years with regular maintenance. Clearly, there will be variations in quality of design and manufacture, and the potential buyer should

Figure 6.1 Gear-driven windpump, South Africa

ensure that they are satisfied with the quality of casting and machining of the windpump components. Due to the more expensive materials and construction techniques, small gear-driven windpumps are likely to be more costly than a simpler design with the same rotor diameter.

In general, manufacture of gear-driven windpumps in the developing world is restricted to the more developed parts of Asia (e.g. India) and South America, where foundry facilities and gear-cutting are more common. However, gearboxes are sometimes imported from Europe, Australia or South America due to lack of indigenous manufacture.

Commercial direct-drive windpumps

A direct-drive windpump is usually similar in appearance to the conventional gear-driven design, but does not use a gearbox. Instead the force is transmitted from the rotor to the pump-rod directly by means of a crank or a cam. This means that there is one stroke of the pump-rod for every turn of the rotor.

The crank mechanism is simpler than a gearbox, and can be manufactured with fairly standard workshop facilities. Crank machines are more suited to village-level operation and maintenance and can provide a sustainable water supply in under-developed regions. Due to the simpler design and construction the cost has the potential to be less than that for a gear-driven machine.

A wide range of quality of design and manufacture is possible with direct-drive windpumps. In the right circumstances, a well-made machine can be very durable and reliable, using modern design techniques along with materials and engineering of a high standard.

Conversely, less rigorous construction techniques and inappropriate design can lead to an unreliable machine with a short lifetime. The

Figure 6.2 Direct drive windpump, Zimbabwe

lifetime of a typical direct-drive machine will be about 10 years.

Direct-drive windpumps are well suited to larger, village-scale applications, as the larger rotor diameter means that the rotation speed will be low. However, problems can develop with smaller rotor diameters, as rotation speeds will be higher causing greater wear on the crank and on the pump itself. A direct-drive windpump is shown in Figure 6.2.

Informally-manufactured windpumps

In certain countries of the developing world there are a small groups of artisans assembling small numbers of windpumps on a casual basis. The designs may be original, or are more likely modified from existing designs to suit locally available components.

Construction materials may include wood (particularly for the tower) and whatever else is to hand, and can be worked with a minimum of facilities. Almost all of these windpumps will be direct-drive machines, using a rudimentary crank or cam system. An example is shown in Figure 6.3.

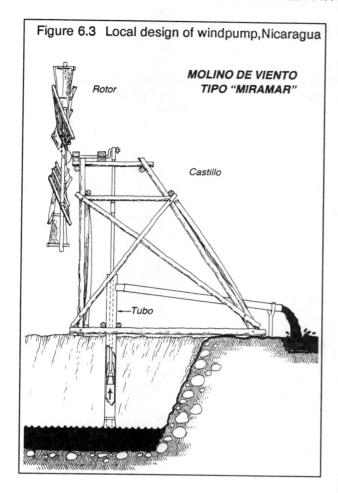

Figure 6.3 Local design of windpump, Nicaragua

MOLINO DE VIENTO
TIPO "MIRAMAR"

Rotor

Castillo

Tubo

Traditional designs are built in villages in South East Asia (China and Thailand) but this is now a decreasing activity as rural electrification and the use of engines spreads.

Clearly, the reliability and performance of these windpumps will be poor in most cases, but the attractions to the user are low costs and easy user maintenance.

It is difficult to say how many such pumps are in use, as the village workshops that make them would not usually market them commercially or sell them outside the local vicinity. Some windpumps may even be constructed by the users themselves as one-off items. Again costs will be low, probably at the expense of reliability and performance.

Locally-manufactured pumps can provide a valuable service in remote or impoverished regions where farmers or villagers have virtually no free capital to invest in a more expensive commercially-made design of windpump.

Wind-electric pumping

In the more developed countries there tend to be more manufacturers of wind-electric generators (see Figure 6.4) than mechanical windpumps and many of them recommend their systems for use with electric motor pumpsets.

A wind generator generally has a smaller number of blades than a mechanical pump (typically 2 or 3, seldom more than 6) and rotates at a much higher speed. The power output of the generator is in the form of direct current at 12 or 24V for the smaller machines and alternating current at higher voltages for the larger ones. Rotor diameters can be anything upwards of about 1m.

Wind-electric systems have several potential advantages. For instance any surplus electrical power may be stored in batteries and used for other purposes such as running lights or a radio. Another useful feature is that the generator does not have to be located directly over the borehole or well, and can be positioned to capture the best wind regime.

Wind-electric systems can be expensive, and to build a reliable generator requires a

Figure 6.4 Wind-electric pumping system

fairly high level of technology. Most commercially-available generators are manufactured in Europe and the US, although there are several models manufactured in-country in the developing world.

In general, wind-electric systems require less routine maintenance than mechanical windpumps. However, when problems occur with the generator or pumpset the technology will be unfamiliar, and spares will have to be imported. Therefore, even where they are available, wind-electric systems will seldom be suitable for use in most developing-country applications.

Unconventional designs

There are also a few types of windpump on the market that fall into none of the categories described above, and involve some unusual aspect in their design, aimed at giving a better performance (e.g. variable

stroke machines), cheaper construction, or some other specific advantage.

An example is the pneumatic transmission windpump. The rotor drives an air compressor, which via a high-pressure hose, drives the pump. Like the wind-electric systems this means that the rotor can be located away from the well, and the transmission involves no moving parts. Until recently such a pump was marketed by Bowjon of the US.

Several companies produce windpumps whose rotor rotates about a vertical axis (see Figure 6.5). Because of the low torques that are produced these machine are not generally used with reciprocating pumps. Instead they use either rotary pumps or 'rope-and-washer' type pumps. These devices are not widely used or manufactured.

Climax of South Africa have used a modified gearbox design on some of their larger gear-driven models to drive a rotary screw-pump. These are designed to perform well at higher windspeeds.

6.2   Survey of Products

The Buyers' Guide

During the winter of 1991/2 all known manufactures of wind-powered water pumping systems were contacted and asked to provide information on their products. The information received is shown in detail for each manufacturer in the format of a buyers guide in Appendix A.

Each entry contains pump specifications in terms of rotor size, number of blades, type of transmission (geared or direct) maximum pumping head and approximate cost. The entries are ordered by continent, and within that alphabetically by manufacturer. A photograph has been included where supplied. A list of addresses of all manufacturers and suppliers known to be still operating is given in Appendix B.

The manufacturers who supplied information are listed in Table 6.1, together with their country of origin and the range of windpump sizes they produce. The last column shows the transmission types produced.

Figure 6.5   Panemone vertical axis windpump, South Africa

Table 6.1 Windpump manufacturers and their capabilities

Company	Country	Rotor Diameter (m)	Drive
AbaChem Engineering Ltd.	UK	2 - 6.3	G
Aermotor	USA	1.8 - 4.8	G
Aureka	India	5.5	D
Auto Spare Industries	India	2 - 5	G
Bergey Windpower Co.	USA	3 - 7	E
Bharat Heavy Electricals Ltd	India	5	D
BJ - Steel	Denmark	3.6	D/E
Cataventos	Brazil	3 - 1.6	G
Climax Windmills	South Africa	2.4 - 5.5	G/D/R
Dempster Industries Inc.	USA	1.8 - 4.2	G
Ets. D. Hermaneau	France	2 - 2.5	D
Energomachexport	Russia	1.2 - 3	G
Essex Associates Inc.	USA	1.8 - 4.8	G
Facogsa Srl.	Peru	4.5	D
FIASA FAB	Argentina	1.8 - 4.8	G
Gaviotas	Columbia	2.0	D
Bobs Harries Engineering Ltd.	Kenya	2.0 - 7.4	D
Industrias Jober Ltda.	Columbia	2.5 - 3	G/D
ISERST	Djibouti	2.7	D
KMP Pump Co.	USA	2.4 - 3	D
LMW Windenergy B.V.	The Netherlands	3 - 5	E
MERIN (PVt.) Ltd.	Pakistan	3.6 - 7	G/D
MidWales Productions	UK	1.8 - 4.8	G
NEPC-MICON Ltd	India	3	G
Ets. Poncelet & Cie	France	1.8 - 2.5	D
Pwani Fabricators	Kenya	3.7 - 4.9	G
Sahara Engineering Co.	Sudan	5	D
Serept Energies Nouvelles	Tunisia	5	D
Sheet Metal Kraft	Zimbabwe	3.6	G
South Africa Plant and Eng. Co.	South Africa	2.3	R
Southern Cross	Australia	1.8 - 7.5	G/D
Star Engineers	Sri Lanka	3	D
Stewarts & Lloyds	Zimbabwe	6	D
Thermax Corporation	USA	0.6 - 2	G/E
Tozzi & Bardi	Italy	4 - 6	D
U.SA Economic Development Co.	Thailand	2.4 - 6	
Vergnet Sa.	France	10	E
Wind Fab (Mayee Eng. Ltd.)	India	3	G
Wire Makers Ltd.	New Zealand	2	D
W.D. Moore Ltd.	Australia	1.8 - 4.2	G

Key to drive types: G = gear-driven, D = direct, E = wind-electric, R = rotary

There are thought to be a great many more manufacturers producing windpumps at the time of writing than responded to the survey, and so this list should by no means be considered exhaustive. One region that is considerably under-represented is South America. The UNDP/World Bank Global Wind Energy Project (GWEP) conducted by the in 1987 positively identified 9 manufacturers in Argentina alone, and several others in each of Columbia, Peru and Bolivia.

Technical aspects

Over half of the windpumps that featured in the survey fall into the category of the standard gear-driven machines, with direct-drive windpumps making up the bulk of the remainder. There are several wind-electric systems in the list, most of which are of European or US manufacture. A handful of the machines on offer fall into the 'unconventional' category, being vertical axis or having a rotary drive.

The geared and direct-drive windpumps almost all have between 12 and 24 blades depending on rotor size, although some have as few as six (the NIVA 3000 from Star Engineers in Sri Lanka) and one machine is made with up to 40 blades (U.SA Economic Development Co. of Thailand).

In terms of physical size, the more conventional gear-driven and direct-drive windpumps vary in rotor diameter from 1.6m for a portable windpump by Cataventos in Brazil, to the 7.5m direct-drive pump manufactured by Southern Cross of Australia. The normal size range available from a typical manufacturer is between 1.8 and 4.8m (6 to 16 feet), and a survey in Argentina suggested that machines with rotors around 1.8 to 2.4m (6 to 8 feet) are by far the most popular.

Maximum-pumping heads vary greatly between different pump types, sizes and manufacturers. It is a figure which depends on both the pump and the windmill, but in general most manufacturers will be able to supply an appropriate combination for any reasonable head.

In practice few systems are able to exceed heads of about 150m. Some of the very smallest machines, or those that do not use reciprocating pumps, may only be capable of lifting water over 10 or so metres.

Costs

Windpump prices can vary greatly, even between machines of similar design. Part of the reason for this is that most manufacturers supply only to the local market, and so prices in different countries can become independent of each other to some extent. However, the main price differences are due to the large variation in manufacturing costs and margins from country to country. In developing countries labour costs will be fairly low, and most of the total cost will be due to materials. The price of the windpump will depend to a large degree on the quantity (and quality) of steel used. Cheaper designs may cut corners on strength, corrosion protection or quality of bearing materials, leading to a less reliable machine.

The graph in Figure 6.6 shows how windpump cost varies with the rotor diameter, and illustrates well the spread of prices. The three different symbols on the graph represent direct-drive, gear-driven and wind-electric pumping systems and the cost is ex-factory including tower and pump. In general the wind-electric systems are the most expensive for a given rotor size.

The mid-range is occupied by gear-driven machines, although there is still quite a spread. The most expensive gear-driven machines are of European or US manufacture. The lower boundary of the gear-driven windpump prices is likely to be set by the cost of building (or importing) a gearbox.

At the lower end of the price range are some of the direct-drive windpumps. Depending on quality and country of origin these reach from within the gear-driven price range, down to just a few hundred dollars ($US). Each pump shown on the graph is described more explicitly in the buyers' guide (Appendix A), and all the data have been taken from manufacturers information during the 1991/2 survey.

From the scatter of the prices, it is clear that it is difficult to define a 'typical' price for a windpump of a given size. However, for the

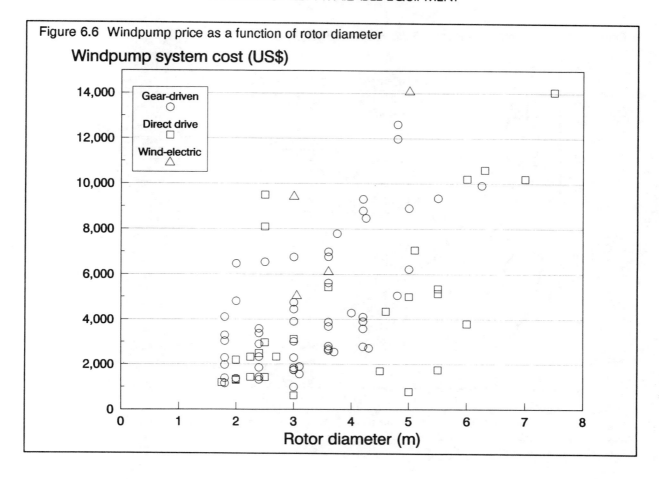

Figure 6.6 Windpump price as a function of rotor diameter

purposes of economic analysis (see Chapter 7) this must sometimes be done. The most general case to consider is that of a standard gear-driven windpump, and it is found that the price is best described as a function of the rotor area rather than its diameter. Using a statistical technique called regression, a figure has been calculated for the cost of a typical gear-driven windpump made in the developing world of 280 US$/m² of rotor area. This figure uses only the price data from Africa, Asia and South America. If only data for gear-driven windpumps manufactured in the more developed regions (i.e. Europe, US, and Australia) are used, a much higher typical cost of 480 US$/m² is found, although it should be noted that Australian prices are generally lower than those in the US. The above figures have been quoted in Chapter 7.

The price data in this section illustrates the variation that can be expected, and it is recommended that real prices from appropriate manufacturers are used if performing calculations on the economics of windpumps.

6.3 Manufacturers' Data

In the process of choosing a windpump the potential buyer will be presented at some point with a manufacturer's brochure, which will contain information on the construction and performance of the windpump. The quantity and quality of information provided vary greatly between different manufacturers, and some of the principal features to look out for are those that are given in the buyers guide entries in Appendix A.

However, the windpump performance (i.e. how much water it will pump) is the most important factor and the hardest to interpret. Information can be presented in many different ways and trying to compare data from different manufacturers can be confusing. Some brochures may contain inaccurate or incomplete data, or may present information in a way that is misleading.

Be suspicious of high output rates where the head and windspeed conditions are not expli-

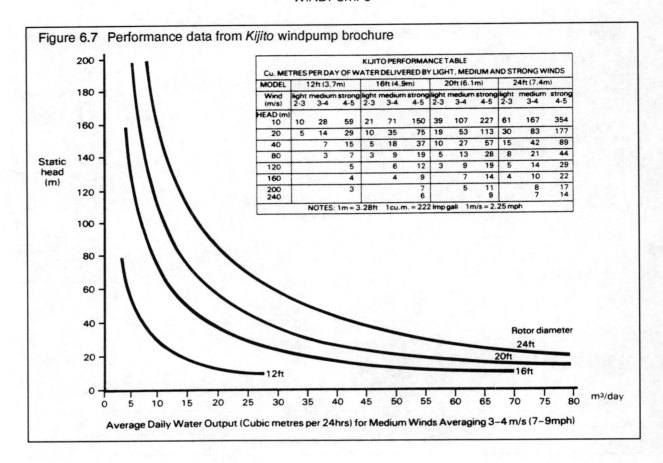

Figure 6.7 Performance data from *Kijito* windpump brochure

KIJITO PERFORMANCE TABLE												
Cu. METRES PER DAY OF WATER DELIVERED BY LIGHT, MEDIUM AND STRONG WINDS												
MODEL	12ft (3.7m)			16ft (4.9m)			20ft (6.1m)			24ft (7.4m)		
Wind (m/s)	light 2-3	medium 3-4	strong 4-5	light 2-3	medium 3-4	strong 4-5	light 2-3	medium 3-4	strong 4-5	light 2-3	medium 3-4	strong 4-5
HEAD (m) 10	10	28	59	21	71	150	39	107	227	61	167	354
20	5	14	29	10	35	75	19	53	113	30	83	177
40		7	15	5	18	37	10	27	57	15	42	89
80		3	7	3	9	19	5	13	28	8	21	44
120			5		6	12	3	9	19	5	14	29
160			4		4	9		7	14	4	10	22
200			3			7		5	11		8	17
240						6			9		7	14
NOTES: 1m = 3.28ft 1cu.m. = 222 Imp gall 1m/s = 2.25 mph												

Rotor diameter
24ft
20ft
16ft
12ft

Static head (m)

Average Daily Water Output (Cubic metres per 24hrs) for Medium Winds Averaging 3–4 m/s (7–9mph)

m³/day

citly specified. Some manufacturers quote outputs at unrealistically high windspeeds, which are likely to be far in excess of what the average site will experience.

The better brochures will give a performance in litres (or gallons) per hour or per day for a range of different windspeeds and pumping heads. An informative way to present performance data is as a graph. It is usual to show output versus pumping head, with several lines representing either different windspeeds or different sizes of rotor.

Another important point to check is that the quoted windspeed is a mean speed and that

the output is not for a brief test with constant windspeed. As discussed in Chapter 4, there can be a large difference between these figures, which depends on the windspeed distribution and the starting and furling windspeeds of the windpump. However, output will usually be quoted at a mean windspeed as this is likely to give a higher figure.

An example of clear and informative presentation of data is shown in Figure 6.7. This is an extract taken from the sales literature of the *Kijito* range of windpumps (based on the I T Windpump) made in Kenya by Bobs Harries Engineering Ltd.

Economics of Windpumps

7.1 General Considerations

Installation and operation of any pumping system requires a long-term financial commitment and the consequences of inadequate assessment beforehand can be dire. In most cases economics will be one of the central factors to consider when choosing a water-lifting system.

Defining what is 'economic'

An economic assessment can take many forms of varying degrees of complexity, but the questions that must be answered are 'what is the cost/value of owning this pump?' and 'do the benefits outweigh the costs?'. Values can be assigned to the various costs and benefits associated with the pump so that they can be compared in numerical terms.

Alternative pumping systems can also be compared with each other, to see which achieves the maximum benefit for the least cost. However, it is also important to realize that not all the relevant considerations can be easily reduced to monetary terms, and the final decision should always be made on a balance of factors.

For a pumping system to be economic in an absolute sense, the value of the benefits must outweigh the costs. In reality this is very difficult to calculate. Although costs can be quite easily quantified, the value of benefits are more subjective. This is particularly true in the case of village water supply, where the main benefit may be access to clean water.

The situation is more straightforward for irrigation pumping or livestock watering, as for a pumping system to be worthwhile, the extra income from the additional crops or animals must outweigh the costs of the pump.

Using economics to compare different types of pump for a given job is simpler, and this is the technique on which this chapter will concentrate. It is assumed that the need for a certain quantity of water exists, and the costs of different pumping methods can be compared to see which one can meet the requirements in the cheapest way. The cost for each can be expressed as a meaningful figure such as a price per unit volume of water, or a cost per family per year.

In many situations, the potential pump user will be constrained by the amount of water available, and so will not have a real choice at all. In the majority of situations in which handpumps are presently used, the wells may not yield enough water for the installation of a mechanized pump. Extraction of more water would quickly pump the well dry, or increase the drawdown to such an extent that pumping is no longer feasible. The potential yield of the well is therefore the primary limiting factor on all forms of mechanized pumping of groundwater.

Pumping options

The chief alternatives to windpumping are diesel and handpumping, and they are the methods that will be considered in this chapter. Solar photovoltaic pumping can also be a competitive technology in remote regions, and its popularity in the developing world continues to grow.

For the renewables (wind and solar) the most important factor is the level of the resource. This is particularly true for windpumping, because the power obtainable for a certain rotor size is roughly proportional to the cube of the windspeed. Therefore, for a given hydraulic energy demand, the size of windpump required depends critically on the mean windspeed that can be expected: a small increase in windspeed would allow a substantially smaller windpump to be used.

Because the cost of a windpump will depend on its size, the economic viability relative to other methods will vary according to the windiness of the site.

Diesel and petrol pumps

Diesel pumpsets are also common throughout the developing world for larger-scale village pumping and irrigation applications. They are usually over-sized for the application, and would be used for an hour or two each day. As they are relatively cheap to buy, accurate sizing is not such a concern as for wind or solar power.

Other factors that are of more concern are the price and availability of fuel and maintenance services. Fuel prices can vary enormously between different countries and even within the same country. At some times diesel fuel may simply not be available, or may be of very poor quality. Diesel pumpsets require regular maintenance and are not generally left to operate unattended for any length of time.

Petrol-driven pumpsets are not generally used because of their relatively short life and the problems of obtaining good-quality fuel in remote areas.

Handpumps

Handpumping is the mainstay of village water supply in the developing world. It is a simple well-established technology that is cheap and generally reliable. However, handpumps cannot deliver large quantities of water, and they require a person to spend valuable time working the pump. In a village where the handpump is becoming over-subscribed, it may well be cheaper in the long-run to

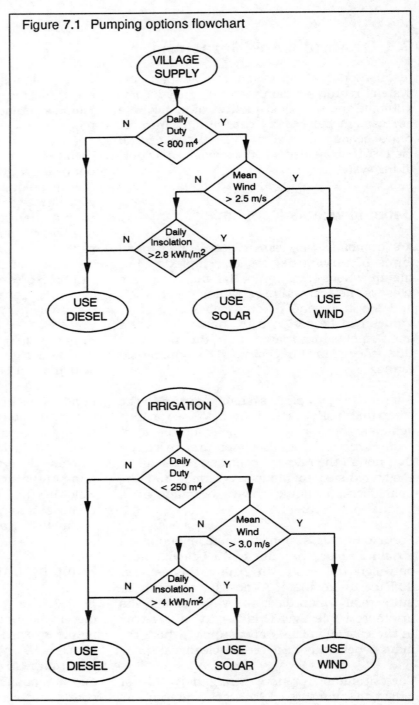

Figure 7.1 Pumping options flowchart

replace it with a windpump than to drill a new borehole for a second handpump.

Handpumps cannot provide quantities of water sufficient for irrigation except on a micro-scale. Unlike diesel, wind or solar powered pumps, handpumping does not usually allow the use of a storage tank or distribution system.

Solar pumps

Solar pumping is also a reliable and simple technology whose use is on the increase in the developing world. Solar panels (called photovoltaic modules) convert the sun's light directly into electrical energy which is used to drive an electric pumpset. The amount of power they can produce is proportional to the area of the modules and the intensity of the sunlight. Therefore for a certain power demand, less sun means that more modules are required, and the system will be more expensive.

So, as for windpumping, the relative economics are dependent on the level of the resource. Solar pumps are expensive to buy initially, but require very little supervision or maintenance and, of course, no fuel.

To summarize the above comments, it can be said that:

- Renewable technologies (wind and solar power) are expensive to buy, but long-lived and cheap to run, needing no fuel, no operator and little maintenance.

- Conventional technologies (hand and diesel) are less costly to buy, but can cost more to run, needing an attendant or operator, frequent maintenance, and fuel for a diesel.

At a very general level a flow chart such as those shown in Figure 7.1 can be used to indicate the conditions in which the different pumping options may be more appropriate. Depending on the hydraulic duty (product of volume and head in $m^3.m$ or m^4), and the solar and wind resources available, the charts lead to either wind, diesel or solar photovoltaic pumping as being most appropriate. Naturally this type of simple chart would never be used as the basis for a decision. As

well as non-economic (e.g. social and institutional) considerations, a site-specific economic analysis is a useful aid to decision making. This is described in the following sections.

7.2 Practical Economic Analysis

There are two ways of looking at the value of any project in qualitative terms. The *economic* approach takes the standpoint of the government, and so considers its value to the economy as a whole. It therefore looks at costs which exclude taxes and subsidies. In contrast, a *financial* assessment is an evaluation from the buyer's or user's point of view. Therefore taxes, subsidies, interest payments on a loan, etc. must all be taken into account.

Only economic assessment will be covered here, since that is the most general case, but it is relatively straightforward to convert this to a financial assessment using appropriate figures for your own situation.

In an economic evaluation, the following parameters are usually considered:

The life-cycle costs: the sum of all the costs of the system over its lifetime, expressed in today's money.

Payback period: the time it takes for the total costs to be 'paid for' by the monetary profits and other benefits of the system.

Rate of return: the magnitude of the profits and benefits expressed as a percentage annual return on the initial investment.

Payback period and rate of return have two disadvantages. Firstly, it is not always easy to express the benefits gained in monetary terms, and secondly, they do not necessarily take account of how long the system will last, nor any future costs that will be incurred as time goes by.

Life-cycle costing is the most complete analysis and is the usual method for determining whether an application is economic. Payback period is often used to give a quick and simple indication of whether an application is likely to be economically viable.

For a life-cycle costing, not just the initial costs, but all future costs for the entire operational life of the pumping system are considered. The period for the analysis must be the lifetime of the longest lived system being compared.

Example

A windpump costs more to buy than a diesel generator, but the windpump should last about 20 years. The diesel generator might last 10 years, using a certain amount of fuel each year. So in this case the analysis period is 20 years. In addition to the capital cost, the cost of a replacement diesel after 10 years, plus 20 years' worth of fuel must also be included for the diesel option. In addition, the costs of maintenance and repair for the two systems over the whole 20 year cycle must be included. Depending on the exact figures, either the windpump or the diesel system will work out cheaper overall.

To make a meaningful comparison, all future costs and benefits have to be discounted to their equivalent value in today's economy, called their 'present worth' or PW (also called present value). To do this, each future cost is multiplied by a discount factor calculated from the discount rate. A discount rate of 10% per year would mean that in real terms it makes no difference to a farmer whether he has $100 now or $110 dollars in one year's time. Therefore a cost of $110 dollars one year from now has a 'present worth' of $100.

The concept of discount rates can be confusing, but the central point is that the further in the future the cost, the lower its present worth, and the less impact it has on the total life-cycle cost. The steps in a life-cycle cost calculation are shown schematically in the flow chart in Figure 7.2.

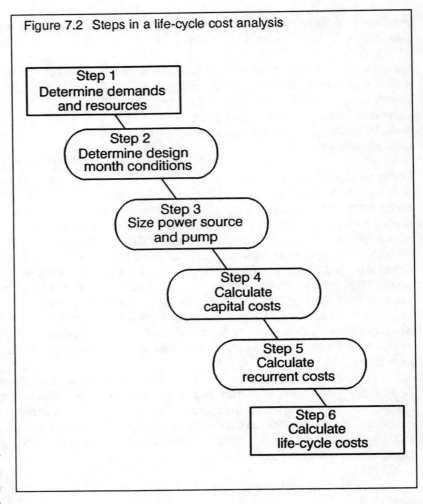

Figure 7.2 Steps in a life-cycle cost analysis

Step 1
Determine demands
and resources

Step 2
Determine design
month conditions

Step 3
Size power source
and pump

Step 4
Calculate
capital costs

Step 5
Calculate
recurrent costs

Step 6
Calculate
life-cycle costs

Economic factors

The calculation of life-cycle costs requires values to be known for the following:

Period of analysis: the lifetime of the longest-lived system under comparison.

Excess inflation (i): the rate of price increase of a component above (or below) the general rate of inflation (this is usually assumed to be zero).

Discount rate (d): the rate (relative to general inflation) at which money would increase in value if invested (typically 8-12%). Alternatively, the annual rate at which future costs and benefits decrease when expressed as their present worth.

Capital cost: the total initial cost of buying and installing the system. This usually includes well-boring, water storage and distribution.

Present Worth calculations

(a) Single payment

For a single future cost **Cr**, payable in **N** years time, the Present Worth is given by:

$$PW = Cr \times Pr$$

Example : It is estimated that a new pump will be required for a certain windpumping system in 10 years time. We will assume that a new pumpset presently costs $500, that pump prices do not change relative to general inflation, and that the discount rate is 10 %. Using Table D.1 with **i** = 0, **d** = 0.10 and **N** = 10 (the number of years hence that the payment is to be made), this gives a discount factor **Pr** of 0.39. Therefore the present worth of this future cost is:

$$PW = \$500 \times 0.39 = \$195$$

(b) Annual payment

For a payment **Ca** occurring annually for a period of **N** years the Present Worth is:

$$PW = Ca \times Pa$$

Example : The fuel costs for a particular diesel generator are $50 per year, and it might be assumed that diesel fuel prices will rise at 4% above inflation. Assuming a discount rate of 10% and a length of analysis of 20 years, Table D.2 (with **i** = 0.04, **d** = 0.10, and **N** = 20) gives a cumulative discount factor **Pa** of 11.69. The present worth of the diesel fuel costs is therefore:

$$PW = \$50 \times 11.69 = \$585$$

Operation and maintenance: the amount spent each year in keeping the system operational, including an attendant if necessary.

Fuel costs: the annual fuel bill.

Replacement costs: the cost of replacing each component at the end of its lifetime.

Calculation of present worth

There are two types of calculation that are used in life-cycle costing when expressing a future cost or benefit as its present worth.

The first is used to calculate the present worth of a single payment, say the replacement of a battery after five years.

The second is used to calculate the total net present worth of a recurring cost, such as annual fuel or maintenance costs. This is the sum of many discounted single payments over the analysis period.

To avoid the need for complex equations, the relevant PW can be found by multiplying the actual cost by a factor that can be found in the tables in Appendix D at the back of this book. The formulae used to calculate the tables are also included.

Life-cycle cost (LCC)

For each payment to be made during the lifetime of the system, the Present Worth can therefore be determined using the discount factors Pr and Pa. The sum of all the Present Worths is the total life-cycle cost of the system. In summary the total life-cycle cost of the pumping system is the sum of the following components:

- Initial equipment and installations costs

- The sum of annually-recurring costs (fuel, operation, maintenance) discounted to their PW

- The sum of all replacement costs (discounted to their PW) over the analysis period

The system with the lowest life-cycle cost can therefore be said to be the most economic system when all future costs are taken into account.

The process could end at this point, but there are two further ways of expressing the life-cycle costs that are more meaningful and easily understood.

Annualized life-cycle cost (ALCC)

This is the total LCC expressed in terms of a cost per year. However the LCC cannot simply be divided by the number of years in the analysis, as this takes no account of the change in value of money due to inflation and discount rates.

The LCC must instead be divided by the factor **Pa** from Table D.2, found using the chosen discount rate, inflation rate of zero, and a number years equal to the analysis period. This is really the reverse process of discounting, and the result is expressed in $/year for each system.

Unit water cost

Probably the most valuable figure for comparing different water pumping systems is the net cost of raising each cubic metre of water for each system.

This can be determined from the annualized life-cycle cost (ALCC) as given in the following equation:

Unit water cost = ALCC / Annual demand

where unit water costs is in $/m³, ALCC is in $/year and annual demand in in m³/year. The total quantity of water supplied each year can be estimated as the sum of all the daily water demands for each month. In some cases the system will may produce more water than is needed in some months. In that case the value used should be the *useful* water delivered.

When comparing two systems, it is often useful to see how the unit water cost varies depending on the size of the system, or to examine the effect of varying the windspeed or the diesel fuel price. Graphs illustrating these variations can be computed by hand, or preferably using a spreadsheet program on a personal computer.

The examples in the following subsections illustrate this process for both village water supply and irrigation pumping for general cases.

The life-cycle costing process may appear rather long and complex, but if a little care is exercised it provides a relatively straightforward way to make a valid economic comparison between different options.

Considered together with other non-economic factors, this forms a vital part of the decision-making process. Of course it should always be remembered that any economic analysis is only an estimate of future conditions, and that it is wise to allow for unforeseen events by choosing a fairly conservative scenario.

Guidance on costs

When doing your own simple life-cycle cost analysis (like that shown on the life-cycle cost sample calculation sheet) it is always best to use real price data. Unfortunately this may not always be available, and the numbers in Table 7.1 are intended to give some rough guidance to help fill in any of the missing information. The figures have been compiled from a database of manufacturers' and users' information, and are relevant to the conventional gear-driven design of windpump. Note that there are two figures for windpump capital cost, one derived from data from developing-world manufacturers and one from data from manufacturers in Europe, the US and Australia.

There are short sections on sizing and costing diesel and hand pumps in Appendix E at the end of this book. For further reading on solar pumps see the *Solar Photovoltaic Products Guide*, listed in the bibliography.

Table 7.1 General windpump data

Capital cost	: Developing nations 280 US$/m² of rotor area.
	: Industrialized nations 480 US$/m² of rotor area.
Pipework	: 5 US$/m
Water storage	: 60 to 150 US$/m³
Borehole drilling	: 60 to 200 US$/m (village supply or livestock)
Well-digging	: 5 to 15 US$/m (irrigation or livestock)
Installation	: 20 % of hardware capital cost
Windmill lifetime	: 20 to 30 years.
Pump lifetime	: 5 to 10 years
Pump cost (Repl.)	: 300 to 800 US$
Maintenance	: 2 to 3 %/year of installed hardware cost
Operating costs	: None

LIFE-CYCLE COSTING CALCULATION SHEET

System Description: *Windpump for village water supply 30m head, 20 m3/day*

Parameters

Period of Analysis **= 20 yrs**   Excess Inflation i **= 0**   Discount Rate d **= 10%**

Capital Cost

Hardware: $.....*10315*.........

Installation: $.......*843*..........

Total: $ *11158*

Operation and Maintenance:

Annual Cost $.......*145*..........per year

Discount Factor (Pa) .....*8.51*..........

Present Worth: $ *1237*

Fuel

Annual Fuel Costs: $.....*Nil*..............per year

Discount Factor (Pa): ....

Present Worth: $ *Nil*

Replacements

Item	Year	Cost	Pr	PW
Pump	*10*	*500*	*0.39*	*195*
			Total $	*195*

Total Life-Cycle Cost $ *12590*

Annualisation Factor (Pa)   ....*8.51*.............

Annualised Life-Cycle Costs $ *1479*   per year

Water pumped per year   ....*7300*............m³

Unit Water Cost $ *0.20* $ / m³

7.3 Economic Analysis for General Cases

The technique of life-cycle costing described above has been used in this section for two very general scenarios: rural village water supply and irrigation.

Rural village water supply is characterized by a fairly even year-round demand, and relatively small quantities of water. This water supply must be clean enough for drinking, and so it is usually pumped up from groundwater a few tens of metres below the surface. Because of the small quantities of water required, and the high value placed on clean drinking water, windpumping for village supply can be cost-competitive given a good wind site.

Irrigation pumping is characterized by the demand for very large quantities of water, but only at specific times of year. This means that for most of the year the pump is idle or oversized. For this reason irrigation pumping is often harder to justify economically. Irrigation water is generally taken from surface water (e.g. canals or water holes) and storage is neither practical nor necessary in most cases.

Livestock watering has many similar characteristics to village water supply. However, capital costs may be lower, as the water source is less critical, and distribution will not be required. Village supply and irrigation have been used in the examples below because they represent the extreme cases.

Costing scenarios

The initial costs for the equipment and costs of replacement parts have been taken from current manufacturers' data wherever possible. The initial costs also include the expense of drilling a well or borehole, building a water storage tank and installing all the equipment.

The annually recurring costs include maintenance of the whole system (pump, water-source, storage tank and distribution system), and fuel and operational costs where appropriate.

The results are shown graphically in Figures 7.3 and 7.4 and each graph shows the way that the unit water cost varies for increasing daily water demand. The general trend is that at low daily demands the cost per m³ is high but falls quickly. At higher daily volumes the curve flattens off to a constant cost per m³.

There are graphs for two different heads for each of the village water supply and irrigation pumping scenarios and each graph shows several curves representing different pumping options. The options that have been compared are windpumping, conventional diesel pumping and handpumping (where feasible), as these are the most common competing technologies.

For windpumping three different scenarios have been calculated, with windspeeds of 2.5, 3 and 4m/s. The curves are cut-off where the rotor diameter needed for such a situation would exceed that of any commercially-available machine. For diesel pumping there are two cases shown, representing the range over which parameters such as fuel price and pumpset costs can vary between different countries.

As handpumps are very low capital cost, the largest component of their life-cycle cost will usually be labour costs. In a village water supply situation the villagers would simply pump their own water as and when they needed it, and so the labour cost is not paid by them directly in monetary terms. However, an *economic* (as opposed to financial) analysis must take account of the total cost to the community, and so it is correct to include the cost of the time spent pumping that might have been spent more productively in other activities. It should be realized that the life-cycle cost depends quite critically on the rate assigned (e.g. so many dollars per day), but it is difficult to make a meaningful estimate of it. In the example US$1/day has been used.

It is also important to realise that these results are very general. Although each line has been shown consisting of discrete points it is more appropriate to think of a range of values either side of the line. Where two lines are fairly close together, the two systems may be considered to be comparable to one another within the accuracy of the analysis.

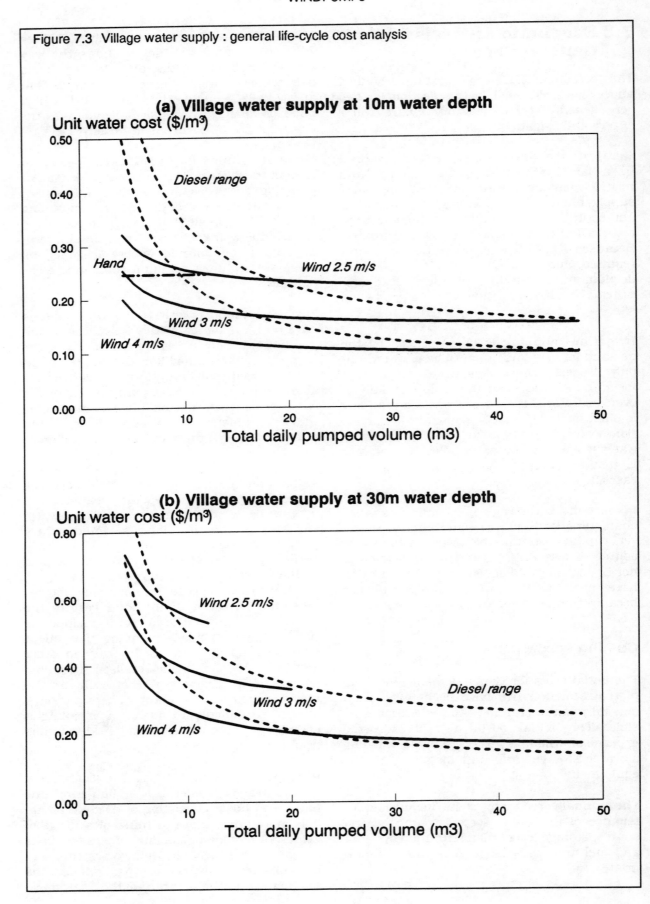

Figure 7.3 Village water supply : general life-cycle cost analysis

(a) Village water supply at 10m water depth

Unit water cost ($/m³)

Diesel range

Hand

Wind 2.5 m/s

Wind 3 m/s

Wind 4 m/s

Total daily pumped volume (m3)

(b) Village water supply at 30m water depth

Unit water cost ($/m³)

Wind 2.5 m/s

Wind 3 m/s

Diesel range

Wind 4 m/s

Total daily pumped volume (m3)

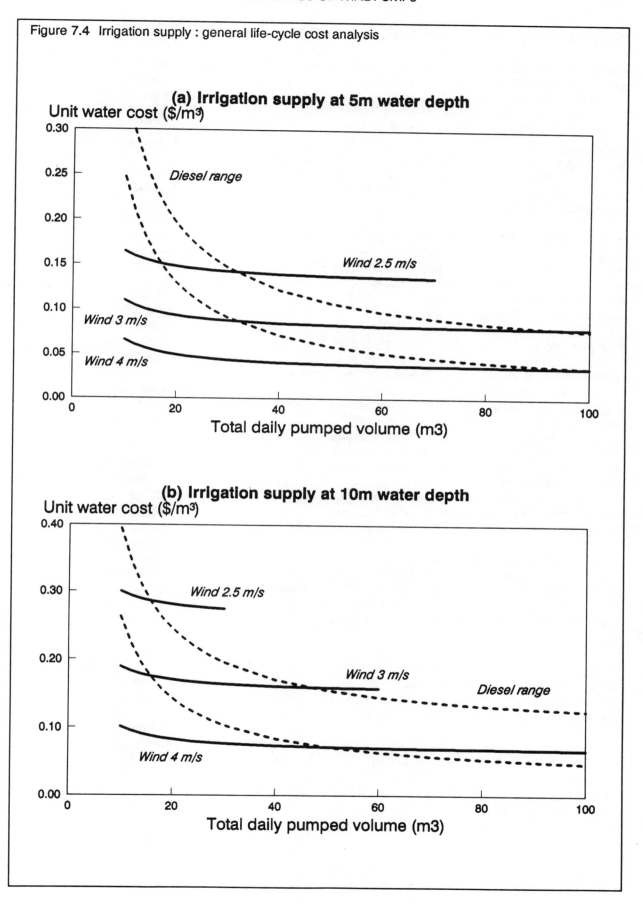

Figure 7.4 Irrigation supply : general life-cycle cost analysis

(a) Irrigation supply at 5m water depth

Unit water cost ($/m³)

Diesel range

Wind 2.5 m/s

Wind 3 m/s

Wind 4 m/s

Total daily pumped volume (m3)

(b) Irrigation supply at 10m water depth

Unit water cost ($/m³)

Wind 2.5 m/s

Wind 3 m/s

Diesel range

Wind 4 m/s

Total daily pumped volume (m3)

Village water supply

The results of a life-cycle cost comparison using conditions specific to a general village pumping scenario are illustrated in Figures 7.3 (a and b), which represent situations with water depths of 10 and 30m respectively. The demand in each graph goes up to $50m^3$ which at a daily demand of $25l/day/person$ corresponds to a village of 2000 people. The windpump cost used is that for a developing country (280 US$/$m^2$).

In the first case with a 10m water depth it is clear that at lower daily volumes, windpumping is the cheaper option. Even at windspeeds of just 2.5m/s, windpumping is cheaper than diesel for daily volumes up to 10 or $20m^3$, depending on the diesel cost scenario.

At higher daily volumes, both the diesel and windpump unit costs tend to level out, and which is the more economic depends largely on the diesel cost scenario chosen and the windspeed. For instance, if the high diesel cost case is chosen, then windpumping at 3 m/s is always more economic, but using the low diesel cost case, it is only more economic up to about $15-20m^3/day$. However, windpumping at windspeeds of 4m/s is more economic than both diesel cases over the whole range of daily volumes, tending towards the lower diesel line at very high deliveries.

Handpumping is a good alternative to diesel at this depth, but still more expensive than windpumping in winds above 2.5m/s. The comments made above concerning the dependence of handpump economics on labour rates should be born in mind.

The second case has a water depth of 30m. Again, windpumping with windspeeds of 4 m/s is cheaper or comparable with diesel over the whole range of demands, and considerably more economic below $15m^3$.

At lower windspeeds (3m/s), windpumping is economic below about 10 to $20m^3/day$ depending on diesel cost scenario. Windpumping at windspeeds of 2.5m/s will only be economic if diesel costs are high and daily demand low.

It is important to note that the unit water costs include the borehole drilling costs.

Irrigation pumping

A similar life-cycle costing was carried out for irrigation pumping, for which the results are shown in Figures 7.4 (a and b) for water depths of 5 and 10m respectively. In this case water demands are much larger (up to $100m^3$ per day) although the heads are lower. Hence pumping would be from a shallow hand-dug well which is far cheaper than a borehole. Because field distribution and application systems tend to be leaky, a volumetric efficiency of 80% has been assumed (i.e. 20% of the pumped water is lost). As before, high and low diesel cost cases are shown.

For the 5m water depth case, the windpump curves show little sensitivity to the daily demand, although there is a factor of two difference between the 3 and 4m/s cases. At windspeeds of 4m/s windpumping is more economic over the whole range of demands, even in the low cost diesel case. At windspeeds of 3m/s windpumping is always the most economic option below $30m^3/day$, and lies below the high cost diesel line over the whole range.

At 10m water depth, windpumping at 4m/s windspeed is always the most economic option for demands up to about $50m^3/day$ and well below the upper diesel limits even above this. However the economics of windpumping at 3m/s windspeed is largely dependent on the diesel cost case used. Windpumping at windspeeds of 2.5m/s would seldom be useful for irrigation.

Summary

From these general cases it is clear that given the right water and wind conditions, windpumping can be more economic than (or at least comparable to) other pumping methods. The graphs also show how the economics of the windpumping option depend critically on the mean windspeed, but that unit costs are less sensitive to demand than for diesel pumping.

Of course, these analyses are dependent on many factors and assumptions, and it is recommended that the user perform his own life-cycle cost comparison using the most accurate local data.

However, the following statements will generally be true:

- Windpumping with mean windspeeds of 4m/s or over will probably be comparable with diesel pumping for most applications

- At very low demands and heads, handpumping may be the best option, depending on the value placed on labour time for pumping

- Irrigation over heads of about 10m will seldom be economic using a windpump.

Note that for the irrigation case above, only the costs have been considered, and not the additional income from the crops. This poses a difficult problem when calculating the economics, as it is no use installing an irrigation system if costs are greater than benefits. Figure 7.5 illustrates schematically the relationship between the water delivery and the annual profitability. Note that if the 'payments' line does not cross the 'additional income' line, then there is no region in which irrigation is viable. The exact figures would of course change with differing crops and climates and local farming knowledge is essential.

The benefits from irrigation can be increased by using a cropping pattern that distributes the demand more evenly throughout the year. A useful mode of operation in drought-prone regions is to use the pump to ensure security of supply of staple crops, while using excess water for more valuable specialized crops. In general it is more difficult to economically justify windpumping for irrigation than for village or livestock water supply due to the large quantities of water required.

Another important factor is security of supply and water storage. For village supply, when diesel and windpumping costs are comparable it is probably advantageous to buy a windpump, and then in periods of calm it is possible to provide supplies from storage. For irrigation, the quantities of water required are likely to be too large for more than minimal storage to be feasible. So even if the unit cost is comparable or even lower than that of diesel, a calm period could have disastrous results for a sensitive crop. It is difficult to build these factors into an economic model (i.e. accounting for lost crops revenues based on the probability of occurrence and duration of calm periods). At this level, such considerations are best left as non-economic factors, and used together with a simple economic analysis to make a qualitative decision.

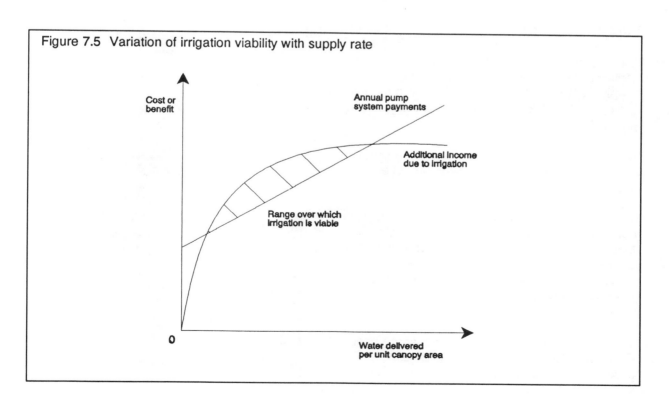

Figure 7.5 Variation of irrigation viability with supply rate

Choosing a Windpump

8

8.1 Introduction

The end result of a site evaluation is to answer the question 'What size of windpump is best suited to the site and what will its capabilities be?'. This must include all aspects from the assessment of water demand and resource availability to the sizing of the windpump itself and the other components of the pumping system.

All windpump manufacturers will use their own sizing methods that are suited to their products. Therefore this chapter simply gives a methodology for an approximate sizing, to give the potential user some idea of the feasibility of windpumping in his situation, and what to expect in terms of hardware requirements.

Supply scenarios

The requirements for the various windpump applications are very different. Village water supply tends to represent one extreme, producing high value, clean water in fairly small quantities.

Irrigation is at the other extreme, pumping large quantities of surface water over low heads.

Meanwhile, livestock watering, which is the most widespread application, lies between these two in many respects. Water sources can be either open wells or boreholes, and heads can vary depending on the conditions. However, daily quantities required tend to be fairly small.

Throughout the chapter there are some parts of the methodology that are common to all three applications, but where differences exists, the cases will be dealt with separately. Figure 8.1 shows an example of a village water supply installation.

There are three main technical factors that set the limits for a given situation:

* The demand for water

* The availability of water

* The wind resource

Figure 8.1 Rural village water supply installation

Demand assessment

Even if there is a good supply of wind, the demand for water and the availability of that water may present serious difficulties. Factors such as seasonally-changing populations and water tables are all unknown quantities that complicate the situation. The concept of demand is not always valid at all, as in many cases the production capacity of the well is simply not great enough. In this case all that can be done, other than drill a deeper well, is to estimate just what fraction of the ideal demand can be met, and decide if this still justifies installation of a windpump.

In this chapter the problems of system selection will be simplified into a few straightforward steps, while still maintaining some degree of realism. Clearly for many communities or donors the question of cost will also be critical, and will itself be a factor affecting demand. For example a lower per capita water supply may be acceptable if the cost of a larger pumping system is too great.

The economic aspect is even more important for livestock watering and irrigation, which can only be justified where the extra income from the crops (or livestock) outweighs the costs of the pumping system. Therefore all systems should be costed as described in Chapter 7, and compared against alternative means of pumping.

8.2   Physical System Layout

The first step in the process is to draw a rough scale plan of the site and decide where the various system components will need to go. This will usually be dictated by the position of the water source and by the local factors affecting the wind pattern.

Water source

For village water supply the source will almost certainly be a hand-dug well or a borehole as surface water is not usually fit to drink. Clearly, if an existing borehole is to be used, then the position of the pump will be dictated by this. However, if a new well is to be dug or drilled, then a position should be chosen that is unobstructed upwind (and preferably downwind). This means avoiding nearby buildings or trees and preferring areas with grass or only low vegetation. For irrigation pumping the water source is more likely to be a canal, water-hole or shallow well, but the same arguments apply. If there is wider scope for the positioning of the pump, then effects of the local topography should also be taken into account. Read Section 4.2 on local wind effects for more advice on siting.

Water storage

For village water supply and livestock watering a storage tank is usually positioned near to the pump, and should only be as high as is necessary to provide the head for the distribution system (if any). A small system will not justify a distribution system, and will simply have a tap (or taps) on or near to the tank. A larger system may need several stand-pipes at different points to allow for the number of users. Remember that the tank and windpump are both very heavy and that soft ground should be avoided. There will probably need to be some compromise on the siting of the pump: on the one hand the distance to walk to get water (or the length of the distribution system) should be kept short; on the other hand the pump should be sited away from the buildings of the village.

Irrigation with windpumps can be difficult, because a very large tank is needed if water is to be stored to cover lull periods. This may be either too expensive or simply too large to be feasible, in which case it will not be possible to grow highly water sensitive crops.

Irrigated area

For irrigation pumping the desired area (in hectares or m²) of the plot to be irrigated should also now be decided upon. One hectare (ha) equals 10,000 m². Most farms in developing countries will be only one or two hectares, and in practice it may not be feasible to irrigate more than this anyway. However, it is more likely that only the slightly more affluent farmers, owning perhaps 4ha, will be able to afford (or raise the credit for) a windpump. In any case, the area that it is wished to irrigate may have to be reconsidered at a later stage in the sizing.

Distribution

For irrigation purposes a tank is not usually feasible, although it is possible to use open pond storage. However, it is more usual to apply the water directly to the crops by some kind of distribution system. The two most practical methods are the low-head drip system and the hose-and-basin method. Both are highly efficient and require a driving head of 1 to 2m. The main difference is the cost. A low-head drip may cost US$2500 per hectare whereas the hose-and-basin method may only cost a few hundred dollars. The method chosen will also depend on the crop grown.

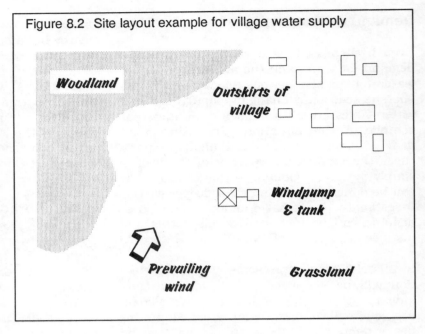

Figure 8.2 Site layout example for village water supply

Woodland

Outskirts of village

Windpump & tank

Prevailing wind

Grassland

For any system the position of its components should be carefully thought out, taking account of the changes in ground level or the slope of the field. A low-head drip or a hose-and-basin system would usually be fed via a small header tank (say 2m above field-level), to filter the water and provide enough head to get an even distribution over the field.

In addition to the cost of the piping itself, it must be remembered that extra head is needed to drive water through longer pipes. This means using extra energy which could have been better used to provide a greater quantity of water. For this reason it is better to site the tank near the pump, so that the run of pipe through which water must be pumped is less.

At the end of this step the user should know the rough layouts of all the components in his system, including the approximate lengths of pipe required, and for irrigation the crop area in m² to be irrigated. He should also know the relative heights of different parts of the system (e.g. tank, field levels etc), as this will be needed to calculate the energy output required of the pump at a later step. Figure 8.2 shows an example of a site layout for a village water supply case, This case will be pursued as an example throughout the chapter.

8.3 Water Resources and Requirements

The typical water sources for village water supply, livestock watering and irrigation pumping are very different, but in all cases it cannot be stressed too strongly that the water resource data is the foundation on which a site evaluation and system sizing is built. The more reliable the data, the more likely it is that the installation will be a success.

Drawdown

The usual source for village water supply will be a borehole. The most likely limiting factor on the amount of water that can be pumped will be the availability and depth of groundwater. Most boreholes or wells drilled for village water supply have only enough capacity to use a handpump (< 2m³ per hour) because they cannot refill fast enough.

As water is pumped from a borehole the water level will drop below that of the surrounding water-table. It is this difference in levels (called the drawdown) that causes water to flow into the well through its walls.

As the pumping rate increases the drawdown also increases until a point is reached where the well is emptied out. This should never happen as the pump seals will be seriously damaged. Therefore a lesser pumping rate must be used which can be sustained.

This means that the in-flow through the wall equals the outflow, with several metres of water still in the bottom of the well. This rate is called the maximum sustainable pumping rate, and will be used as an upper limit for designing the rest of the system.

The windpump must be sized such that the maximum sustainable pumping rate is not exceeded, even in the windiest conditions.

The exact relationship between the drawdown and the pumping rate will depend on the diameter and total depth of the borehole or well, and also on the permeability of the rock or soil. The way the water-table is 'drawn down' around a borehole is illustrated in Figure 8.3.

Geology

In situations where the water-table is found in shallow sandy soil, say near a river or a lake, the water can be regarded as submerged surface water. The abundant nearby supply and high permeability of the soil will mean that the drawdown will usually be relatively small.

Where boreholes must be sunk into deep aquifers below caps of impermeable rock, the drawdown will be very much larger for the same pumping rate, This can form a signifi-cant part of the total head, and can even equal the static head.

So without including drawdown in the sizing, there is a danger of under-sizing the wind-pump as well as over-sizing, as the real level in the well once pumping begins may be well below the static water-table.

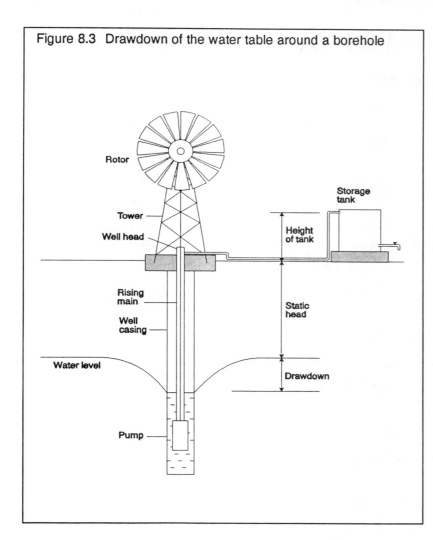

Figure 8.3 Drawdown of the water table around a borehole

Rotor

Tower

Well head

Storage tank

Height of tank

Rising main

Well casing

Static head

Water level

Drawdown

Pump

Shallow wells

In the case of irrigation pump-ing the water source is most likely to be close to the sur-face.

This is because irrigation demands large quantities of water, and to pump over-high heads would require too much power to be economic.

Secondly, shallow water sources are less likely to be limited by their capacity than deep boreholes. Typical sources for irrigation wind-pumping may be canals, rivers, land-drains or other surface water, or shallow wells or waterholes.

As refill rates are fast, draw-down does not generally pre-sent a problem, although some care should still be taken with shallow wells. Even in good alluvium a relatively small system can induce a drawdown of 3 or4 m in a high-yield open well.

Livestock

Cattle are not as sensitive to water quality as humans and the water source for livestock watering may be from either surface water or wells. However, pollution of surface water by cattle while drinking can be a problem, and it may be preferable to pump the surface water some distance or to dig a well.

Seasonal variations

Seasonal changes in the depth of the static water level should also be taken into account where information exists. They will be more pronounced in shallow water situations, and depths can vary by several metres in some places. This can sometimes double the head of shallow irrigation pumping systems. The level in deep aquifers tends to remain more constant. For practical purposes monthly data are desirable.

Existing hydrological data

Although it is never possible to exactly predict the drawdown relationship before drilling, some idea of its extent can be gained from data from neighbouring wells that have been measured regularly over a number of years.

If an existing well is being used there may be useful information in the driller's records (which are often logged by the government). When a borehole is drilled a test should be performed by using a diesel pump to pump out water at a certain rate and measuring the drawdown when a steady state has been reached. The maximum safe rate and the corresponding drawdown should be recorded.

UNDP, UNICEF and other agency-sponsored-groundwater studies have now been conducted and documented for most parts of the world and it is very rare to find an area with no groundwater survey reports at all.

Summary of initial site data

Source	Borehole
Total well depth	50 m
Inside diameter	6 inches (15.2 cm)
Test pumping rate	6.5 m³/hour
Static water level	14.0 m
Level at end of 8 hrs pumping	30.0 m

So in this example the drawdown at the test rate is : 30.0 - 14.0 = 16.0 m.

Depth to static water level:

Jan 19.0	Feb 19.5	Mar 20.0	Apr 21.0	May 18.5	Jun 16.0
Jul 14.0	Aug 13.5	Sep 14.0	Oct 15.5	Nov 17.0	Dec 19.0

Length of pipe (pump to tank)	15 m
Length of pipe (tank to taps)	0 m
Number of taps	3
Height of tank (well-head to tank top)	2 m

Initial site data

At this point it may be useful to summarize the information that has been collected so far, including both the site layout data and the water resource data.

Together, these will be used in a later step to calculate the total head over which water must be pumped. The data collected for the example village water pumping system is shown in the box on the previous page.

From the data above it is possible to deduce several things:

- The maximum drawdown will occur at the maximum sustainable pumping rate

- The windpumping system that is chosen should not be able to exceed the maximum test pumping rate (i.e. $6m^3$/hour) at the given head, whatever the wind conditions. This will prevent oversizing

- The borehole must be deep enough to accommodate the maximum drawdown at the lowest static level with a few metres left in the bottom, i.e. 16m drawdown plus the April static depth 21m gives 37m, so a 50m borehole is quite adequate, providing it penetrates the aquifer sufficiently.

8.4 Water Demand

Village water supply and livestock watering

Domestic water requirements vary markedly in response to the actual quantity of water available. For example, the average domestic consumption in western Europe and the USA is between 150 and 250 litres per day.

At the other end of the scale the level in rural areas of the developing world varies between about 5 and 35 litres per day. In drought conditions many people survive near the biological minimum of just 2 litres of water per day.

Factors affecting demand

Community population statistics are seldom recent or accurate, and populations change over time and with the seasons. This is particularly a problem in rural Africa; in Latin America and Asia populations and their demands tend to be more stable. In addition, per capita requirements change from season to season, Some local knowledge of the population and their way of life is therefore an essential requirement in estimating the water demand for a village.

With windpumps that exist solely for livestock watering, the animal population will at least be controlled, but there will still be seasonal changes in demand. During the dry season it may be necessary to make provision for extra livestock watering.

When sizing a windpump it is important to remain flexible concerning the water demand. If in just one month the pump cannot match the criteria set, then people will usually adapt to the conditions for a short period; this will be more preferable to them than having no pump at all. Similarly, if more water is being pumped than required for a certain month, they will be happy to use it.

The most likely constraint on the available supply will be maximum extraction rate of the well, and at some times of year it may not be possible to produce the ideal desired daily volume.

Per capita requirements

A WHO survey showed that the average per capita water consumption in developing countries varies widely, with houses with a piped supply using up to five times more than if water must be carried from a public water point.

The WHO subsequently defined 40 litres/day as a short-term goal for the developing world. However, in many parts of Africa 10l/day is regarded as an acceptable and realistic quantity for rural areas. This covers drinking water and cooking needs.

Low-tech mechanical systems (e.g. hand-powered pumps) may produce in the range of

Figure 8.4 Animal-powered pump

20 to 25 litres per capita per day, and it is recommended that this be used as a rough figure in calculating windpumping demand figures if local requirements are not directly known.

Seasonal changes in per capita consumption (for both people and animals) may be about 15 % either side of the mean, with the maximum being in the dry season. If no local data on seasonal water use is available this can be used to adjust the monthly demand.

Seasonal changes in village population may be more marked. As other water sources dry up, people that were using them may come from further afield to use the windpump.

For example, a survey of handpumps in a region of Ghana found that the average number of people using the pumps increased from 750 to 1250 in the dry season. Again, local knowledge is a great help in making assumptions of this kind.

Livestock requirements

Table 8.1 shows typical per capita daily water requirements for a range of livestock. Calculation of water demand is simpler in the case of livestock watering, as the number of animals using the pump throughout the year will be known quite accurately. Figure 8.5 shows an example of a windpump supplying water for livestock.

Collating demand data

The population and livestock data throughout the year can be collected into a table giving the minimum daily village requirements for each month.

An example is shown in Table 8.2. The first column shows the number of people needing water each month, and the second column shows their per capita requirement in litres. (Although this cannot be known to an accuracy of one litre, a smooth variation may be included to simulate seasonal changes.)

The third column is the ideal total daily demand in m^3/day and is simply the first two columns multiplied together and divided by 1000 (to convert from litres to m^3). In the example, April is the driest month and August the wettest.

This total demand is used in the next step in which the windpump is sized.

Expected losses of water in the distribution and storage system should also be accounted for. However, distribution efficiencies for village water supply systems should be very high (above 95%) and so water losses can probably be ignored.

Table 8.1 Typical daily water requirements for livestock

Animal	Water Requirement (Litres per day)
Horses	50
Dairy Cattle	40
Steers	20
Pigs	20
Sheep	5
Goat	5
Poultry	0.1

Figure 8.5 Livestock-watering installation

Table 8.2 Example calculation of ideal daily demand for village water supply			
Month	Number of people	Per capita demand (l/day)	Daily demand (m³)
Jan	550	26	14
Feb	575	27	15
Mar	600	28	17
Apr	625	29	18
May	550	27	15
Jun	500	25	13
Jul	425	23	10
Aug	375	21	8
Sep	400	22	9
Oct	425	23	10
Nov	450	24	11
Dec	500	25	13

Irrigation pumping

To find the daily volume of water that the pump must produce, it is necessary to estimate the crop water requirement, which is traditionally defined in m³ per ha per day or mm of water per day. This is perhaps the most difficult part of the evaluation, as the crop water requirement depends on many factors.

Firstly it will vary between different crops, and its annual distribution will of course depend on the cropping pattern. At some points of its growing season a crop will need more water than at others. This is most likely to be when it is growing at its fastest.

In addition, crops with a larger canopy (i.e. more or larger leaves) will lose water more quickly by evaporation. This is properly called evapotranspiration, and will also be affected by windspeed and the humidity of the air. The type and condition of the soil will also affect its ability to hold water in the root zone.

All this must be balanced against the rainfall that can be expected in each month, with the irrigated requirement making up the shortfall. In many tropical areas (e.g. monsoon regions) the rainfall distribution can be predicted quite reliably, and data will be available from the meteorological office, universities or agricultural institutions. A farmer will know from experience in which months to expect rain and when little or no irrigation will be needed.

Cropping patterns

To make most effective use of the windpump it will probably be advisable (and is usually necessary anyway) to grow more than one crop per year. Then the demand on the pump is more even all year round. For instance a cereal crop could be alternated with a vegetable crop. It should be clear that all these factors require in-depth local knowledge, and that generalizations cannot be made with any accuracy.

Also when using a windpump it should be remembered that crops need to be hardy varieties and tolerant of water shortfall.

Crop water requirements

The fullest method of crop water calculation is given in *Crop Water Requirements* published by FAO (see bibliography).

However, rules of thumb have been developed from traditional irrigation experience, some of which can be applied to windpumping to get a feel for the quantities involved. One of the most useful rules of this type has come out of work on drip systems, and is given as follows:

- In hot dry climates use 7-8 litres/day per m^2 of crop canopy

- In cooler or more humid places 5-6 litres/day per m^2 of canopy is sufficient.

This can be converted to an equivalent in m^3/day per ha of crop canopy, which is more convenient to use in our sizing calculations. Simply multiply the original figure by 10: for instance, $8l/day/m^2$ equals $80m^3/day/ha$.

So if a crop at a certain stage in its growth covers 50% of the ground area, then a one hectare plot will have a total canopy of 0.5 ha.

Therefore (assuming the hot dry regime) the volume required per day would be 80 x 0.5 = $40m^3$ on the field for each hectare of crop.

Clearly, the fraction of canopy cover will increase as the crop grows, and will depend on how closely the crop has been planted. However, this is something that can be estimated with only a little local knowledge of the crop and the way it is planted, and does not need detailed technical information or calculations.

Application efficiency

To find the actual daily pumped volume required the conveyance efficiency of the application system must also be taken into account. For instance 80% efficiency means that 20% of the water is lost before it gets to the field. It is inadvisable to use systems with efficiencies less than 80-90 %.

Calculating pumped volume

The pumped volume required per day can be easily calculated from the rule below:

$$\text{Pumped volume } (m^3/day) = \frac{\text{Plot size (ha)} \times \text{Canopy fraction} \times \text{crop water demand } (m^3/day/ha)}{\text{Application efficiency}}$$

Alternatively this can also be done graphically using the nomogram in Figure 8.6. Both of the methods yield the same result and the choice of which one to use is a matter of personal preference.

Graphical method

To use the nomogram, start on the left-hand horizontal axis labelled 'size of plot' and select the appropriate number of hectares to be irrigated.

Trace a line vertically up to the appropriate canopy fraction line and then right to the 'canopy area' axis. Extend this line further right across to the relevant water requirement line (in l/day/m2 canopy) and then trace a line downwards through the 'volume at field' axis to the appropriate application efficiency line.

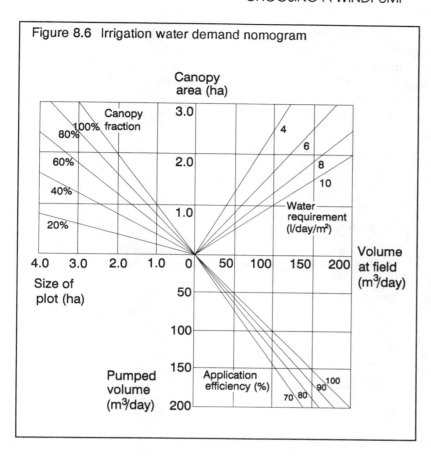

Figure 8.6 Irrigation water demand nomogram

Finally, trace a line left to the lower vertical axis to read off the necessary pumped volume in m³/day. So the line is traced clockwise around the nomogram.

Alternatively it can be used in reverse (anti-clockwise) to find the size in hectares of irrigable land given a certain pumped volume, application efficiency, crop water requirement and canopy fraction.

Definitions

It should be noted that the nomenclature used in this book may not be universal throughout the windpumping world.

For instance, the crop water demand will often be called the 'reference crop evapotranspiration' and expressed in terms of mm/day. This is an equivalent water depth, and is equal to litres/day/m². That is to say, that a litre of water distributed over a square metre would be 1 mm deep.

It is also usual to use a 'crop growth coefficient', which is defined slightly differently to the simple canopy cover fraction used here, but can be taken as the same figure for most practical purposes.

Example
The method that has been described in this chapter is suitable as an initial rough guide and an example is shown in Table 8.3 for a 1.5 ha plot in a very hot, dry region.

It is assumed that two crops per year will be sown: An estimate of the canopy cover

Table 8.3 Example calculation of irrigation requirements

Month	Canopy fraction	Total canopy ha	Crop demand m³/day/ha	Volume at field m³/day	Volume pumped m³/day
Jan	0.1	0.15	80	12	14
Feb	0.2	0.30	80	24	28
Mar	0.5	0.75	80	60	71
Apr	0.7	1.05	80	84	99
May	0.8	1.20	80	96	113
June	0.0	0.00	0	0	0
July	0.1	0.15	60	9	11
Aug	0.2	0.30	60	18	21
Sep	0.3	0.45	60	27	32
Oct	0.4	0.60	60	36	42
Nov	0.5	0.75	60	45	53
Dec	0.5	0.75	80	60	71

fraction has been made and a crop water requirement has been assumed of 8l/day/m² canopy (i.e. 80m³/day/ha canopy) in the hot season (Dec to Jun) and 6l/day/m² (60m³/day/ha) in the cool season (Jul to Nov). Using a low-head drip system an application efficiency of 0.85 is assumed.

An example of a windpump being used for irrigation is shown in Figure 8.7.

Figure 8.7 Irrigation pumping in Ethiopia

8.5 Sizing the Windpump

The sizing method given here is much simplified from a real situation, but serves to give a rough idea of the required pump size. The result of the sizing is to find out an approximate rotor area (or diameter) and a fixed efficiency is used for the rotor/transmission/pump system. In reality the system must be optimized to provide the right combination of pump and rotor for the conditions.

The consequences of different rotor size/pump diameter combinations are discussed qualitatively at the end of the section.

The size of the rotor that is needed to meet a given duty is dependent on the required hydraulic energy and the wind regime. The hydraulic energy requirement is proportional to both the total pumped head and the quantity of water pumped per day.

Design month

The sizing is carried out for each month (or season) separately. The month which needs the largest rotor area is called the 'design month'. This is so called because it is the 'worst case' and represents the extreme conditions that the design must meet.

If the system can meet the requirement in this month then it can, by definition, meet them in every other month.

So the three main parameters that are needed for each month (or season) to size the system are:

● The total pumped head (m)

● Daily pumped volume (m³)

● Expected mean wind speed (m/s)

Method

A logical way to proceed is to create a table such as that in Table 8.4, with the months listed in the far left-hand column. This has been filled in for the village water supply example continued from the preceding sections.

Head

In the second column the total pumped head (m) should be entered. This is the head that is 'felt' at the pump and is the sum of:

● The depth of the static water level below the surface

● The maximum expected drawdown

● The height of the tank (if any) above the surface

● Additional dynamic head due to friction losses

Table 8.4 Example rotor sizing for village water supply

Month	Total head (m)	Volume requ. (m³/day)	Mean windspeed (m/s)	Rotor area (m²)	Rotor diameter (m)	Volume pumped (m³/day)
Jan	37.0	14	3.3	16.0	4.5	14
Feb	37.5	15	3.8	12.0	3.9	21
Mar	38.0	17	4.2	9.9	3.6	27
Apr	39.0	18	4.4	9.4	3.5	31
May	36.5	15	3.5	15.0	4.4	17
Jun	34.0	13	3.2	15.0	4.4	14
Jul	30.0	10	3.3	9.5	3.5	17
Aug	29.5	8	3.1	9.0	3.4	14
Sep	30.0	9	3.4	7.8	3.2	18
Oct	31.5	10	3.6	7.7	3.1	21
Nov	33.0	11	3.5	9.6	3.5	18
Dec	37.0	13	3.4	14.0	4.2	15

The tank height, static water level for each month and maximum expected drawdown have already been found and summarized in Section 8.3 on water resources.

The exact increase in water depth due to drawdown will not be known for a specific pumping rate, and so for safety the maximum value should be used.

The dynamic head is the extra pressure that the pump feels due to friction and turbulence in the pipework. The narrower and longer the pipes the greater this component will be. If the flow rate and pipe diameter is known then the additional head can be calculated as described in Section 9.1. This will also vary with the material of which the pipe is constructed.

For the sizing it should be assumed that the pipes are large enough that the dynamic head will be small compared to the total head, and so zero can be used. The total pumped head is therefore just the sum of the tank height, static level and maximum drawdown.

Daily volume required
In the third column of Table 8.4 the daily water requirement (m³/day) is entered. This was calculated in the previous section, 8.4 on water demand. Note that this is the demand at the pump, and so must account for any inefficiencies or leaks in the distribution system.

Windspeed
In the fourth column should be entered the mean windspeed for each month. If possible this should be monthly means measured at the hub height of the windpump, or at least 10m above ground level.

In reality, there may neither be the time nor resources to do full measurements, and data from nearby meteorological stations will be needed. Read Chapter 4 for more advice on sources of windspeed data and performing measurements.

For the purposes of the example in this chaper some monthly mean windspeeds have been made up and entered in column 4.

Rotor size

Having filled in the three crucial parameters, the rotor can be sized. This can be done either by calculation or graphically. Both give the same result and are based upon the approximate relationship that was discussed in Chapter 5:

Mean hydraulic power =
0.1 x (mean windspeed)^3 x (rotor area)

where power is in watts, windspeed is in m/s and rotor area in m^2. If you don't have access to a calculator or you simply prefer not to use maths then you will probably want to use the graphical method.

By calculation
The rotor area in m^2 that would be needed in a certain month is:

$$\text{Rotor area} = \frac{\text{(total head) x (daily volume) x 1000}}{\text{(mean windspeed)}^3 \text{ x 367 x 0.1 x 24}}$$

where head, volume and windspeed are taken from columns 2, 3 and 4 of the table and are in units of m, m^3/day and m/s respectively. The rotor area should then be entered in column 5 for each month.

The rotor area is a useful quantity when sizing and costing because windpump costs tend to vary with the rotor area rather than with its diameter. However, windpumps are usually specified in terms of their rotor diameter and to find this from the area is another simple calculation:

Diameter = 2 x $\sqrt{(\text{Area}/\pi)}$ where π = 3.14

Enter the rotor diameter for each month in column 6.

Graphical method
This uses a diagram (properly called a nomogram) from which results can be found by tracing lines in certain directions and reading the results where they intersect the axes.

The nomogram in Figure 8.8 is very simple to use and has been designed to find rotor area and diameter from the three parameters in Table 8.4, namely total head, required daily volume and mean windspeed.

Looking at Figure 8.8, find the figure for daily pumped volume on the lower vertical axis. Trace a horizontal line across to the appropriate 'head' line, and then vertically up to the line corresponding to the mean windspeed for the month.

From that point trace a horizontal line left to the upper vertical axis and read off the rotor area in m^2. Enter this in column 5 of Table 8.4.

Extend the line further to the left to the point where it intersects the curve and then vertically downwards to read off the corresponding rotor diameter from the horizontal axis. Enter this in column 6.

If the figure you want to use for head or windspeed falls between two lines, then simply estimate where the appropriate line would lie and use that. It is important to remember that this sizing method can only be approximate and great accuracy is not necessary.

The nomogram shows an example line for the month of March from the example.

Size selection

Having found the rotor size that would be necessary to meet the requirements for each month, a decision needs to be made of the actual size to choose.

As a first step it is simplest to choose the largest area (or diameter) found, as it is certain that a windpump of this size will be able to meet the demand all year round. In the example this is a rotor of diameter 4.5m.

It is worth bearing in mind at this point that windpumps are not available with diameters of greater than about 8m.

If the rotor size needed for one or more months exceeds this diameter, then it will be necessary either to reconsider the daily volumes for those months, consider other forms of mechanical pumping, or use more than one windpump.

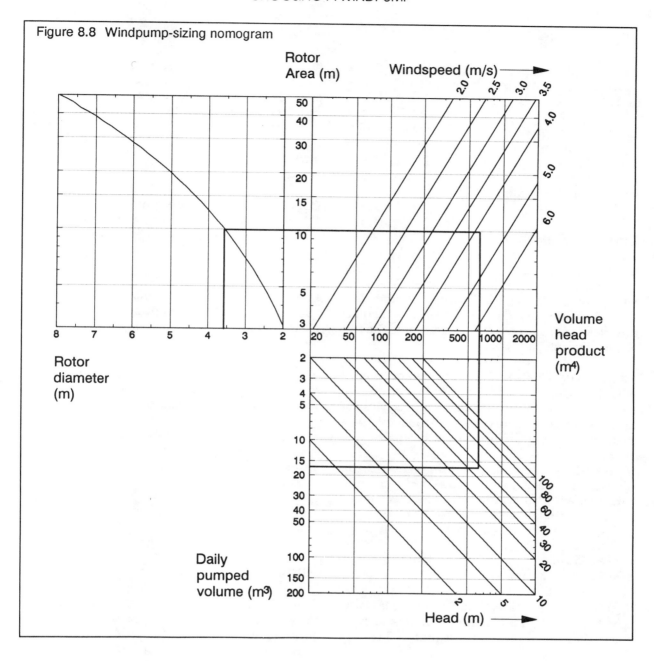

Figure 8.8 Windpump-sizing nomogram

Actual volume pumped

Having chosen a provisional diameter, the actual daily volume of water that will be produced can be calculated for each month.

Put another way, this is like going backwards through the sizing process, using the monthly windspeeds and heads with a constant rotor diameter to find the daily pumped volume.

Again this can be found by calculation or by graph. The result should be entered in column 7 of Table 8.4.

By calculation the formula to use is:

Actual volume pumped =

$$\frac{(\text{rotor area}) \times (\text{windspeed})^3 \times 24 \times 0.1 \times 367}{(\text{total head}) \times 1000}$$

Graphically, simply use the nomogram in Figure 8.8 and trace round in the opposite direction (anti-clockwise).

Starting at the rotor diameter axis, use the windspeed and head for each month, to find the estimated daily pumped volume.

Water deficits

You may also wish to form an eighth column showing the water excess or deficit for each month.

If there are just one or two months that require a slightly larger rotor diameter, the pump will be oversized for the rest of the year. However, the cost of oversizing a windpump will usually be small.

Economic factors

One of the major factors that will influence the choice of rotor diameter is the cost of the windpump. By looking through manufacturers' catalogues for machines of a similar size it should be possible to get some idea of the likely cost.

Alternatively Chapter 7 gives a general way to calculate the cost from the rotor area. In all cases it is worth performing a life-cycle cost analysis as shown in Chapter 7 and comparing windpumping against other methods.

Nevertheless, there may be a maximum capital cost imposed for financial reasons, and this may limit the windpump size. If economic or financial considerations are critical it will be worthwhile checking the economics of different rotor sizes and using the method above to find the expected daily pumped volume in each month.

If this is less than the demand that was specified for certain months, it is necessary to check that the quantity is adequate.

If the only feasible windpump provides far less water than is required in several or all months, then it would be necessary to think carefully about whether windpumping is worthwhile at all. Alternatively a second windpump installation could be considered.

Technical factors

There are other technical factors concerning the matching of rotor and pump sizes that may affect reliability of water supply for the system. The choice will depend on the water resource conditions and wind regime:

The pump bore will affect the performance of the windpump in different circumstances. The smaller the size of pump, the lower the windspeed that is needed to start it. Therefore a smaller pump will allow more frequent operation and hence improve supply reliability in less-windy periods. However, because a smaller bore pump will deliver less water per stroke, the overall water output may be reduced.

Conversely, a large bore pump will only start at higher windspeeds, but because more water is pumped per stroke, it may produce more water overall.

Hence pump sizing demands a compromise between applying a light enough load so that the pump will start in light winds, and having a large enough bore to produce adequate overall output.

The furling windspeed will also affect the way the pump reacts at high windspeeds, and so affect the water output.

So the quantity of water that is actually produced for a given rotor diameter is not fixed, but will vary depending on the way the rotor/pump match reacts to the windspeed distribution. To look at this in detail is beyond the scope of this book, and is a job for the manufacturer when he prepares a quote.

Pipework

One of the final jobs is to size the pipework from the pump to the tank. This should be large enough that at the peak flow rate the dynamic head is less than 10% of the total head, and less than 5% at the flow rate for the mean windspeed. Remember that it is cheaper to buy a larger pipe than a larger windpump. Excess dynamic head can also seriously damage a windpump.

The peak flow rate (e.g. in litres/second) will vary for different windpumps, depending on the rotor/pump matching and furling speed as described above. The peak rate is therefore best taken from the manufacturer's figures once a windpump has been decided upon. The method and the pipe-sizing nomogram from Section 9.1 can be used to ensure that pipe is chosen with a large enough diameter.

8.6 Procurement

Procurement is the process of contacting manufacturers with your requirements, assessing the merit of their replies and finally buying a windpump. In the last subsection the approximate rotor size range was found, and if necessary the water demand modified to fit the conditions. Using that size as a guide, the potential buyer can now begin writing to suppliers to find quotes and specifications for individual machines.

The buyers' guide in Appendix A at the back of this book contains entries for many of the world's foremost windpump manufacturers. There is also a list of addresses in Appendix B. This list is not exhaustive, and the buyer may know of other manufacturers, perhaps more locally situated.

The first step is to write to, or visit, several different suppliers, especially local ones, with a brief description of the requirements. Most suppliers will then send a detailed questionnaire concerning the information they need. One of the most important aspects of corresponding with windpump suppliers, is to *give them as much information as possible* about the site. The better they know the requirements and conditions of a particular site, the more accurate their chosen equipment will be.

It is also important to only specify the site conditions such as water resource and demand data and wind regime (if known): *do not* express a preference for any particular piece of equipment or rotor size. Otherwise the suppliers will not be liable should the installation not perform as required.

Preliminary evaluation

Each reply should be checked to ensure that :

- the windpump offered is complete and includes spare parts, tools and installation and operating instructions

- the windpump offered can be delivered within the time specified

- an appropriate warranty can be provided.

Detailed assessment

A windpump is a big investment. Make sure that all your questions have been answered to your satisfaction before buying a particular machine.

A detailed assessment of each offer should be made with respect to each of the following four areas:

1. Compliance with specification: Check that the information provided about the estimated performance of the windpump matches your requirements and conditions. Also compare them with your own approximate calculations.

Check the windspeed at which any predictions have been made and check that the starting and furling speeds are consistent with your wind regime. Check that the peak pumping rate does not exceed the well's maximum sustainable capacity.

2. System design: The general suitability of the windpump should be assessed taking into account operation and maintenance requirements, complexity of the transmission and other parts, safety features etc.

The lifetime of the equipment should be considered, giving special attention to parts susceptible to wear, e.g. bearings, pump-rods, blade fixings, water-tight seals etc.

Look at the amount of supporting information provided including general assembly drawings and performance information. This should include head versus flow curves for a range of windspeeds, starting and stopping windspeeds and furling windspeed.

3. Capital costs: Compare the various replies, allowing for differences in specifications between models. How does this compare with your own estimate?

4. Overall credibility: The relevant experience and resources of the tenderer should be considered. The tenderer should be able to provide a repair and spares service within a reasonable time if problems with the windpump arise. A warranty should be provided that should ideally cover at least five years.

Implementation

9

9.1 Installation

Having assessed the economics of different water lifting devices (Chapter 7), concluded that windpumping is the most suitable option, chosen the optimum site (Chapter 4) and tendered for and procured the windpump (Chapter 8), there are a number of steps that should be carried out on receipt of the windpump, before installation begins:

- ensure that all the necessary documentation is included, this will generally consist of assembly and installation instructions, a list of parts and spares, a maintenance and repair manual, and a warranty agreement;

- check against the parts list or shipping details that all parts are correct, that none are missing or have been damaged in transit;

- if there is a query on any item always contact the manufacturer before commencing installation.

The installation process can be fairly complicated and time-consuming and care is required in its execution. Before beginning read the *safety* instructions in Section 9.3 and in the manufacturer's handbook. If the manufacturers are not installing the windpump then they should provide detailed installation instructions, e.g. the Aermotor Windmill Corporation include a booklet entitled *Installation, Operation and Maintenance Instructions* (Figure 9.1) with all their machines. However, instructions can vary considerably in quality from supplier to supplier. Poorly-installed machines can easily lead to inferior operation and maintenance problems, hence it is advised that potential windpump buyers view these instructions before making a commitment to purchase. In any case easy installation and maintenance should be one of the key criteria for selecting

a particular design, so the quality of the instructions and the complexity (or simplicity) of the installation deserve careful consideration.

Installation pertains to all the different components that make up a windpumping system, the foundations, borehole or well, rising main, and the windpump structure itself, etc. The information that follows merely outlines the considerations that should be taken into account when setting up a windpumping system and are not exhaustive guidelines. Manufacturer's instructions should be adhered to in all cases and any further queries should be referred back to the manufacturer, to ensure that the terms and conditions of the warranty are not breached.

Figure 9.1 The *Aermotor* booklet on installation, operation and maintenance.

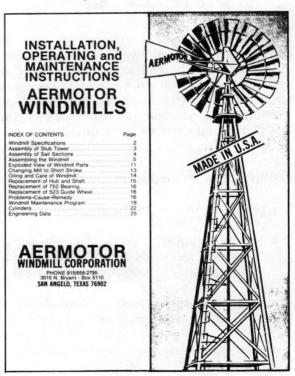

Borehole/well construction

Windpumps are most commonly installed over a borehole or alternatively over an open well.

Borehole drilling is a specialized task which should ideally be assigned to trained professionals with the correct equipment, hence this publication only outlines the procedure and items which should be taken into consideration for operation and basic maintenance (usually minimal) carried out by the owners/users of the system.

The most important factor in construction is the borehole alignment. A slight deviation from the vertical will be particularly significant with a deep borehole. The windpump pump rods must hang vertically or they will wear and, in extreme cases, cause malfunction of the machine. If the borehole is constructed out of alignment this can easily occur.

Boreholes used for wind powered water pumping are usually constructed with diameters of 100 to 150mm (4 to 6"). A typical borehole is illustrated in Figure 9.2. In some cases, e.g. in hard rock strata, boreholes need not be lined.

Plain lining, or screening, is used for the top 10 to 30m to prevent surface water contamination and also through non-water-bearing strata, e.g. clay, or material likely to collapse. In water bearing strata the lining is perforated with 10 to 25mm diameter holes at 50 to 100mm spacing, or with slots 3 to 12mm wide and 150mm long with centres 100 to 150mm apart. The drilled borehole diameter should be larger than the outer diameter of the lining. The resulting space surrounding the lining is back-filled with gravel.

The borehole should penetrate the water table sufficiently to allow water to percolate through the lining as fast as the water is pumped out, even at the highest pumping rates and in the driest seasons, and to ensure that the pump is always submerged, even at maximum drawdown. The gravel in-fill between the drilled hole and the lining must be thick enough to prevent sand and silt from choking the borehole and pump and the

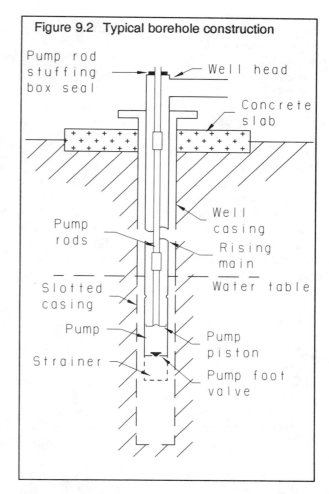

Figure 9.2 Typical borehole construction

Pump rod stuffing box seal — Well head — Concrete slab — Pump rods — Well casing — Rising main — Slotted casing — Water table — Pump — Pump piston — Strainer — Pump foot valve

lining material should be carefully chosen. Lining material is usually chosen from stainless steel, ideal but expensive; steel or galvanised iron, which can corrode; plastics, cheap and lightweight but may deform under load, especially when long; and burnt-clay screens, which are often brittle.

Wells are usually simpler and cheaper to construct than boreholes, as they can be dug by hand. However they are made, they must be lined where they pass through loose sub-soil, either traditionally with masonry or with pre-cast concrete. Wells can be sunk up to 60m deep, usually with a diameter of between two and four metres. As with boreholes they must penetrate the aquifer to provide sufficient submerged surface area. It is difficult to hand-dig to a substantial depth below the water table, but there are other ways of increasing the submerged surface area, e.g. by digging horizontally below the aquifer. If a well is to be used for water for human consumption, care should be taken to ensure that it does not become contaminated.

Windpumps can usually be mounted directly over a limited size well. This size should be indicated on a foundation drawing, in the installation instructions or can be obtained from the manufacturer.

If the windpump needs to be located over a larger open well then there are two installation options. It is usual to install a steel *I-section* beam (RSJ) across the well to carry one side of the windpump tower. Alternatively a reinforced concrete raft, with an access hole, can be cast to cover the well. This is integrated into the windpump foundations. A suitable drawing should be available from the windpump manufacturer particularizing their chosen option.

Borehole/well testing

After a borehole or well is first completed the drawdown should be tested. This is usually achieved using an engine-powered submersible pump. It is necessary because the drawdown can sometimes double the depth of borehole required and indicate the need for a well with more surface area below the aquifer.

The measured drawdown distance should stabilize an hour or two after beginning the test, but this may not occur if the pumping rate is too high. Further information on the aquifer characteristics, in particular the highest sustainable pumping rate, can be obtained by two-hour step drawdown tests; the initial rest-level of the water is measured then pumping commences, at a low rate. The drawdown is measured frequently until two hours later the pumping rate is increased and the process is repeated. When the drawdown fails to stabilize over a two hour period, the pumping rate is too high. Ideally the peak pumping rate for the operational system should be considerably less than this value.

In most countries it is mandatory to provide a record of a borehole, its depth, static water level, and drawdown at a specified pumping rate. The specific country requirements should be ascertained at an early stage to ensure that the correct details are measured before the windpump is installed over the borehole. Borehole *specific yield* is sometimes used to compare the output of different boreholes. This is measured in metres cubed per day per metre of drawdown, hence to calculate this figure the pumping rate (m³/ day) is divided by the drawdown (m).

In both newly-dug boreholes and wells the water should also be tested for salinity and acidity to ensure that it is suitable for drinking or irrigation. A high acidity (pH) or salinity reading should be double checked before a decision is made as to the suitability of the borehole location, especially if the water is for human consumption. The International Standards (WHO) for Drinking Water are illustrated in Table 9.1. Requirements for irrigation water are less stringent, although high salinity levels can only be tolerated on well drained permeable soils.

Foundations

The windpump manufacturer should provide technical information and instructions concerning the type and dimensions of the foundations. Most frequently these are made up of concrete footings, one for each leg of the tower, encapsulated in a reinforced concrete raft around the mouth of the borehole, or across the well shaft.

A large windpump will require up to a cubic meter of concrete for each footing, and more for the raft (the foundations must be massive enough to withstand the forces on the windpump in storm conditions).

It may be possible to mount the smallest windpumps on lengths of steel angle section driven deep into the ground, but even if the installation instructions allow this option it should only be used where the ground is firm, but not too rocky.

Table 9.1	International Standards for Drinking Water		
Property	Unit	Permissible	Excessive
Acidity	pH value	7.0 - 8.5	< 6.5 > 9.2
Salinity	milligrams per litre (mg/l)	500	1,000

Concrete foundations are established by excavating holes around the well head, according to the manufacturer's instructions/drawings. The foundation or *holding-down* bolts are then suspended in the holes using carefully positioned and levelled wooden templates (support is usually provided by wooden stakes driven into the ground) and the suitable concrete mix is poured in around them. The windpump tower, once in place, is then bolted down to the foundation bolts.

A practice sometimes used for small windpumps, but not recommended, is for the tower to be assembled and suspended with its feet in the foundation holes. Considerable care is then required to ensure accurate alignment before the concrete is poured in around the legs (approximately half a meter of the tower legs should be encased in concrete). This method has the advantage of requiring fewer machined components for the feet of the tower. However, the disadvantages of having a fixed windpump tower are such that this practice is best avoided. Difficulties can also arise when servicing the below ground components of the system because the tower cannot be moved.

It should be noted that:

- it is important that the centre of the windpump is accurately located above the borehole. This will only be achieved by correctly locating the foundations, according to the manufacturer's instructions;

- if the windpump is to be laid on its side for erection or repair sufficient space must be available at the appropriate site;

- the concrete, whether surrounding a borehole or well, should slope away from the mouth, forming a graded concrete apron, to prevent surface water entering and causing contamination. This should also channel water away from the foundations. In extreme cases undermining of the foundations can occur by water washing away the subsoil surrounding the well-head;

- the concrete must be kept damp and covered with polythene sheeting, hessian, or some other suitable material, whilst it is curing (reaching the required strength and durability), usually for three to six days as indicated in Table 9.2, otherwise it will crack and be weakened.

Pump rods and rising main

The pump rods, coupling the windpump mechanism to the pump, are a crucial component of the system. The most vital requirement for their installation is that the pump rod joiners are sufficiently tightened, but can still be readily disconnected when raising the pump for servicing. A disconnected or broken pump rod can be catastrophic since the lower part, attached to the pump, falls back into

Table 9.2 Curing time for concrete		
Ambient conditions	Average surface temperature of concrete after casting	
	5°C to 10°C	Above 10°C
Poor	6-10*	4-7
Average	4-6	3-4
Good	2	2

* Curing time measured in days
Note 1. Temperature of the concrete when pouring should not exceed 30°C
Note 2. Ambient conditions after casting are:
    Good: damp and protected (relative humidity >80%, completely protected form sun & wind)
    Average: intermediate between good & poor
    Poor: dry or unprotected (relative humidity <50%, not protected from sun & wind)

the rising main and is usually extremely difficult to recover.

In this context it should be noted that when using wooden pump rods it is usual to use wrought iron couplings. These may be fitted with a quick-release mechanism to help speed up the process of raising and lowering them in the well.

The rising main is required to convey the water from the pump to the surface. The pump is attached to the rising main, and the rising main to the borehole casing at the well head.

No further fixing is usually required because the rising main is heavy due to the weight of water enclosed within it and this effectively inhibits any movement.

The rising main usually consists of standard galvanized steel water pipe in 6m lengths. The installation process is:

- a tripod sheerlegs, or similar equipment, is set up incorporating a chain block and tackle sufficient to carry the weight of the rising main as it is lowered into the borehole;

- the pump, or a part of it (see later), is fitted (including the strainer if required) to the lowest section of rising main;

- a clamp is bolted to the top of this length of rising main, and, using the block and tackle, this section is lowered down the borehole;

- once the clamp rests on the top of the well head the block and tackle can be released and used to raise the next section of rising main vertically above the first, after a clamp has been fitted to the top of this new length;

- the two sections are joined and, after ensuring that the joint is sufficiently tightened, the lifting tackle is used to raise them slightly to remove the lower clamp;

- both sections are then lowered into the borehole until the second clamp rests on the well head;

- this process is repeated, fitting and clamping lengths of rising main, until the pump is sufficiently submerged to ensure that drawdown never allows the water level to come below the pump intake (see next page if a non-extractable pump piston is to be used);

- The extractable pump piston is attached to the pump rods which are lowered down the rising main, section by section;

- in the final stage the well-head fitting is screwed to the top section of rising main, lowered into the casing and tightened.

This process will be modified according to the type of pump installed. If an extractable pump is used, as shown in Figure 9.3, then the above method will be used. The footvalve would be fitted to the lowest section of rising main before the rising main was lowered, but the piston and associated seals would be lowered into place, already fitted to the pump

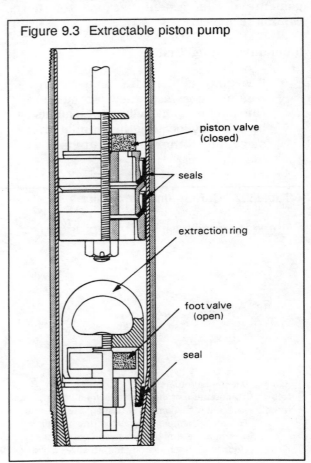

Figure 9.3 Extractable piston pump

piston valve (closed)

seals

extraction ring

foot valve (open)

seal

rod, after the rising main was secured. This involves fitting the piston to the lowest pump rod, and repeating the process of clamping and lowering for the pump rods.

If a non-extractable pump is used, as shown in Figure 9.4, then the whole pump is attached to the lowest section of rising main and, as the rising main is lowered, the pump rods (usually the same length as the rising main sections) are joined.

Finally the length of pumprod needs to be adjusted to ensure that the piston does not quite touch the footvalve at the bottom of its stroke once connected to the windpump. This is usually carried out by coupling the windpump to the pump rods with the windpump crank at bottom dead centre (lowest point of the stroke) and the piston resting on the footvalve. There will then be some means of adjustment provided to shorten the pump rod and raise the piston around 25mm (one inch) above the footvalve. The actual clearance should be as specified by the manufacturer.

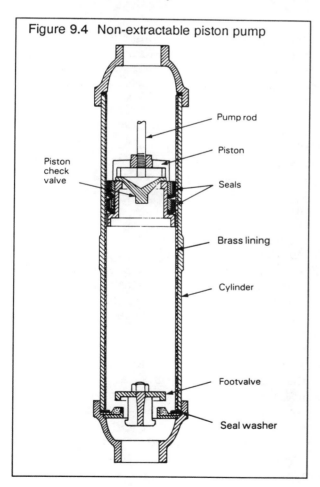

Figure 9.4 Non-extractable piston pump

Pump rod

Piston

Piston check valve

Seals

Brass lining

Cylinder

Footvalve

Seal washer

Tower and rotor

Most windpump towers are made from hot-dip galvanized or painted steel components, but whatever the material, it is constructed by bolting the parts together. A large inventory of nuts and bolts is called for with traditional designs. Parts supplied are simply bolted together according to the manufacturer's instructions to form a three or four-legged tower. It is important to note that all bolts should be securely tightened. If cross-bracing is used then it must be carefully tensioned. The tensioning bolts are designed to permit this to be easily carried out.

Some manufacturers sell an adapter tower top which allows their machine to be installed on an existing or prefabricated tower. If an existing tower structure is to be used it should be carefully examined for signs of corrosion or failure, especially around the tower feet and foundations.

Many traditional designs of windpump are built from the ground upwards. This usually means that a jib with a chain block or crane will be required to raise parts into place, particularly for the larger machines. Some windpumps (especially smaller machines) can be assembled almost horizontal then lifted using either a winch with an 'A' frame, a crane or a drilling rig jib.

A few machines have hinged footings and are specially designed to be assembled horizontal then raised, rather than lifted, with an 'A' frame and winch (foundation drawings should specify a location for the winch fixing). These machines can also be lowered to the assembly position for major overhauls. A windpump with hinged footings being raised after assembly is shown in Figure 9.5. This method of installation is quicker and easier than building from the ground upwards.

It cannot be over emphazised that it is critical to erect the machine precisely so the pump rod is both vertical and accurately centred over the borehole. Some windpumps only have a small tolerance of pump rod misalignment and this is a common cause of failure.

The rotor is invariably fabricated from hot-dip galvanized mild steel components. The blades are rolled from thin steel plate, which is

subsequently gal-
vanized, or from alu-
minium. Typical
windpump rotor
assembly often con-
sist of two concentric
rings of flat bar atta-
ched to 'A'-shaped
spokes. The cam-
bered plate blades
are then bolted to
brackets attached to
the concentric rings.
Again an important
requirement is that
all the rotor nuts
and bolts are
securely tightened
on assembly.

Figure 9.5 Winching a windpump into the upright position in Kenya

Storage tank

Excepting some irrigation applications, or
where windpumps are used for specialized
tasks, e.g. aerating fish ponds or pumping sea
water to produce sea salt, a storage tank is
essential. The tank should be as close as is
reasonably possible to minimize the length of
the delivery pipe and the dynamic head
(discussed in more detail later) that this
implies.

Storage tanks are required to ensure a supply
of water throughout periods of little or no
wind. The storage capacity needed is dependent on the local wind regime and the
importance of maintaining a continuous
water supply.

Typically, storage tanks are sized to incorporate three to five days storage capacity. In
areas where there are periods of extended
calm or where a reasonably high guarantee of
water availability is important, a larger capacity tank is used. Conversely in areas with
consistent winds, e.g. coastal regions or
where an occasional lack of water can be
tolerated, or where an alternative source can
be used in an emergency, then a smaller
capacity tank could be considered. However,
for low-head irrigation applications, where
large volumes of water are handled, it may
not be economic to provide more than one or
two days storage (the cost of large storage
tanks can be a significant proportion of the

total cost) and even this may not be feasible.
Sometimes the soil can store enough mois-
ture to compensate for calm periods.

Different water tank construction methods,
their advantages and disadvantages, are considered in Table 9.3. The water usage will
also play a role in the determination of the
type of tank employed.

Many systems employ steel tanks mounted at
ground level. However, for any system incorporating water supply to several outlets it is
normal practice to use a steel tank mounted
on a stand. This allows the supply to the
outlets to be gravity fed.

A higher tank will be able to feed water
through a greater distance but it is usually
more expensive the higher it is mounted. The
height also adds to the static head through
which the water must be lifted and so a
windpump with a larger diameter rotor may
be needed.

The least costly option is to mount a larger
diameter, shallow tank at ground level,
although this only permits gravity feeding of
water from the tank a short distance to low
water points, such as cattle and sheep
troughs. Ground mounted tanks can be made
from steel, commonly curved corrugated steel
sheets placed on a concrete footing (see

Table 9.3 Storage tank construction methods: advantages and disadvantages

Construction method	Advantages	Disadvantages
Prefabricated plastic	Easy to construct Leak-tight Corrosion resistant	Small sizes only Expensive Slow deterioration in sunlight Prone to rodent & termite attack
Prefabricated galvanised iron	Easy Leak-tight	Small sizes only Galvanising wears through
Site-assembled steel plate	All sizes possible Robust Leak-tight	Expensive Requires painting or lining to prevent corrosion
Site-assembled GRP plate	All sizes possible Corrosion resistant Leak-tight	Expensive Slow deterioration in sunlight
Stone, brick, concrete or soil	All sizes possible Cheap Materials locally available	Waterproof lining required Liable to leak Large tanks need buttresses/ walls to support water weight

Figure 9.6), ferro-cement or simply soil. In the last case an area is evacuated and the soil removed is used to build the sides before the structure is lined with a water proof material, e.g. a butyl rubber membrane. The latter is probably one of the least cost options for very large water storages at ground level.

The choice of tank is best discussed with the windpump supplier, who will be able to advise on the most suitable storage tank knowing the likely output of his windpump in the given wind regime. Some suppliers offer storage tanks, usually steel, whether ground mounted or on a stand, as part of a complete windpumping system.

Water delivery

If the storage tank can be located with the discharge at a lower level than the top of the riser pipe around the pump rod then there is no need to seal the pump rod at the well head. However the storage tank frequently needs to be raised above the top of the riser pipe. In this situation it is necessary to provide a seal where the pump rod emerges from the riser pipe. This is because gravity will always encourage the water to flow to the lowest available outlet. If the delivery pipe is

Figure 9.6 Corrugated steel sheet storage tank for village water supply in Botswana

higher than the top of the riser pipe then the water, acting under the force of gravity, will find it easier to flow out of the top of the riser pipe. A seal will not allow this to occur, hence the water will find another outlet; it will flow down the delivery pipe.

Various types of seal are available from windpump suppliers; the most common are:

- **packing seal:** previously asbestos a suitable substitute is now used. This type of seal depends on being tightly compressed around the pump rod. It is inefficient since it imposes friction on the shaft. It often leaks a small amount and requires regular maintenance to compensate for wear;

- **top-hat seal:** usually rubber, plastic or leather, this seal is located over the top of the riser pipe and around the pump rod. It imposes less friction but once it does start to leak it must be replaced;

- **counter-piston:** this method employs a piston of smaller diameter than the main pump (usually half) to pump some water by displacement on the down stroke of the windpump. It is located in the riser pipe above the exit to the delivery pipe, This can improve the output, but the windpump must be able to tolerate the reverse loads it implies.

In general packing and top-hat seals are most frequently used but are not entirely effective and wear easily. Pumping water through significant heads above the riser pipe should thus be avoided.

As mentioned in Chapter 8 the flow of water through the delivery pipe has an associated dynamic head due to friction between the water and the pipe walls and turbulent flow. The dynamic head is the effective extra height that the water must be lifted through to overcome this friction. It increases as:

- the length of the pipe increases;

- the flow rate of the water increases;

- the roughness of the inside walls of the pipe increases;

- the diameter of the pipe is reduced.

Consequently a shorter pipe length, with a larger diameter, a smooth inside surface and a slower flow rate is preferred to reduce the dynamic head. In practice the flow rate is usually fixed by the size of the system and the length of the pipework by the required location of the storage tank (which should be close to the windpump if possible), hence the only sizing variables are the type of pipework and the pipe diameter.

The delivery pipework should be chosen and sized such that the dynamic head is not greater than 10% of the total head at the peak flow rate, and ideally less than 5%. PVC pipe is most commonly used although it degrades in sunlight over a period of time. Another option is galvanized iron pipe, although this has higher friction losses due to the rougher surface finish. The chart in Figure 9.7, for PVC pipe, can be used to indicate the pipe size required for a specific flow rate.

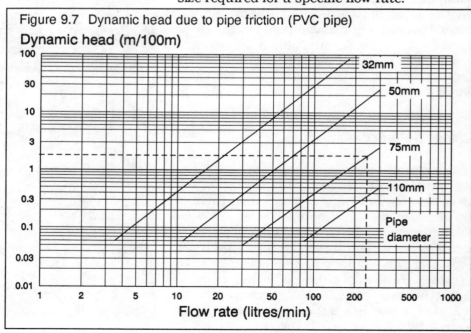

Figure 9.7 Dynamic head due to pipe friction (PVC pipe)

Dynamic head (m/100m)

Flow rate (litres/min)

Using the example village system in Chapter 8:

Maximum pumping rate 6m³/hour (250litres/min)

If an available pipe diameter of 75mm is assumed then on the horizontal axis of the chart draw a vertical line through a flow rate of 250litres/min. From the point where this line intersects the angled line representing the 75mm diameter pipe, draw a horizontal line to the vertical axis, as shown. The point on the vertical axis then represents the dynamic head, in this case 2m/100m of static head. This is merely 2% of the static head.

Maximum static head

Height of tank	2m
Max depth to static water level	21m
Drawdown	16m
TOTAL	39m

$$\text{Dynamic head} = \frac{2 \times 39}{100} = 0.78m$$

Thus for the example in Chapter 8, the dynamic head was assumed to be negligible and this was a valid assumption. For this example it would even be possible to use a pipe with a slightly smaller diameter. However, it should be noted that bends and valves etc. increase the frictional effects so that a margin of error should be allowed to compensate.

The problem of dynamic head is most critical in situations of low static head and high flow rates. In these cases the dynamic head can become a significant proportion of the total head if care is not taken to choose a pipeline of sufficiently large diameter. It is also important to note that when a reasonably long pipeline is connected to a windpump (though not desirable it is sometimes necessary) the machine can be badly overloaded if the wind speed increases. This occurs because when the wind speed increases the water flow rate rises, increasing the dynamic head and hence the total head. This illustrates that it is crucial to evaluate the dynamic head at the peak flow rate.

Another consideration when installing pipeline for windpumping systems is the surge induced by the pulsating nature of piston pumps. This surge is known as water hammer and arises from a pressure shock wave, like that created if a valve or tap is closed suddenly. It usually manifests itself as a loud bang as the water is brought to a standstill, but in extreme cases water hammer can actually burst steel pipes.

Safeguards against this condition are only necessary if the delivery pipeline is long. The most common is the **air chamber** or **air vessel,** another option uses hydraulics in a system similar to car suspension (see Figure 9.8) but this is relatively expensive. The air chamber works on the principle that the air in the chamber is compressible so that if the pressure in the pipeline increases the air is compressed and more water will enter the air chamber, as the pressure falls so the water is expelled. This removes the shock of water hammer and smoothes the flow.

Air chambers are generally essential when using a direct-drive windpump, even with short-delivery pipelines, due to the relatively rapid action of the pump. With conventional gear-driven machines it is less critical and for delivery line lengths less than 30m an air chamber should not be necessary. All air chambers should have some means of replenishing lost air (the air is inclined to dissolve in the water over a period of time), e.g. a bleed valve.

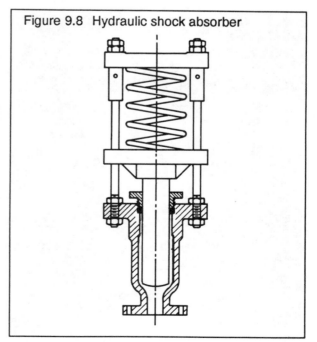

Figure 9.8 Hydraulic shock absorber

9.2 Operation, Maintenance and Repairs

Commissioning

Before the windpump operation commences it is judicious to carry out basic implementation checks and commission the windpump. Wherever possible it should be confirmed that all moving parts are free to rotate/reciprocate as required.

In many cases the furling mechanism tension spring will need to be adjusted to ensure furling at the correct wind speed. It is always better to choose a lighter setting initially so that the windpump furls at lower wind speeds to avoid damaging the new machine. It can always be tensioned once the machine has been seen to run and furl effectively. The operation of the manual furling and the brake (if one exists) should also be checked. Figure 9.9 illustrates the manual furling system being operated.

It is important that any faults are reported promtly to the manufacturer. Most problems should be covered under the windpump warranty, as long as this has not been contravened by poor installation.

Operation

Efficient and trouble-free operation can be achieved if regular maintenance of a high standard is carried out. Manufacturer's instructions make recommendations for maintenance, usually periodic oiling/oil changes and/or greasing. These recommendations are dependent on the transmission and bearing types (although *rolling element bearings* are sometimes greased and sealed for life) and it is wise to adhere to them, not just throughout the warranty period, but for the whole life of the windpump.

In addition to the manufacturer's recommendations it is prudent to consider that:

- the machine is designed to withstand severe situations. However, *fatigue* problems, due to continued cyclical circumstances (the repetition of strong and then gentle wind conditions) faced by all windpumps can lead to cracks or in extreme cases, failures of components. Consequently it is recommended that machines are inspected regularly, particularly at welded joints, bolt holes, etc. for signs of cracking;

- a correctly-installed and well-maintained windpump should operate fairly quietly. If there is any unusual noise when inspecting the machine investigate immediately;

- the windpump must be re-painted periodically to prevent corrosion. If unchecked, corrosion can rapidly destroy the machine in a hostile environment;

- the pump will require occasional replacement seals and piston cup leathers (this operation is considered in more detail later in this section). Unless otherwise specified in the manufacturer's instructions this operation should be carried out every year. If a pump rod seal is fitted at the well head this should be replaced at the same time;

Figure 9.9 Checking the manual furling mechanism on a windpump in Pakistan

- the water carrying components, delivery pipe, storage tank, etc. need to be examined for leaks. A small leak of only one drop per second amounts to a loss of five litres of water over one day;

- air in air chambers can be gradually dissolved in the water, the bleed valve should be opened periodically to allow the air to be topped up;

- after a major overhaul or repair, or the installation of a new machine, nuts and bolts may work loose. These must be checked and re-tightened where necessary (any that are dangerously loose may be noticeable if they rattle).

Maintenance and repairs

It will nearly always be necessary to climb to the top of the tower in order to service the windpump, (see Figure 9.10). Before doing so the windpump should always be disabled by activating the manual furling mechanism and applying the brake, if available. For any operation which entails climbing the tower a safety line is required. A canvas belt (strong enough to carry the weight of its wearer) attached to a short steel cable which can, in turn, be attached to a convenient point on the machine is sufficient although professional belts can be obtained if required.

One potential source of failure is the furling mechanism. As automatic unfurling occurs, when the wind drops to a predetermined safe level (approximately 5 to 6m/s), the spring can violently pull the rotor back into the wind. In itself this does not present a problem, but to ensure that the rotor does not overshoot the unfurled position some kind of stop is employed. It is the force with which the rotor hits this stop that can cause damage. This is especially true when the wind speed falls suddenly.

Many designs of windpumps now incorporate a spring buffer, or similar device, to absorb the impact. This is a most desirable feature for any windpump. The furling mechanism should also be *fail-safe*. This means that if a critical component fails the rotor will simply furl at a very low wind speed. This is the case for a traditional-style furling system. If the spring fails (they are prone to rust and eventually break) the furling system will operate at a much lower wind speed than otherwise intended.

In order to service the pump and replace the seals etc. it must be removed from the borehole. The removal process is the reverse of that detailed for installation. The advantages of using an extractable pump become evident at this stage. The piston and the footvalve can be removed from the borehole without disturbing the rising main.

Figure 9.10 A safe working platform, as on this *Southern Cross* windpump, is beneficial for carrying our regular maintenance

The piston will typically slide 2000 to 3000km a year. It is not surprising then that the seals need replacing every one to two years.

Certain seals made from synthetic leather (used in Australia) are claimed to last up to five years under favourable circumstances, i.e. in clean, silt-free water.

The footvalve will also occasionally

require attention. If the footvalve leaks then the performance of the windpump will decline because the rising main will drain down whenever there is a lull in the wind. This will take some useful energy to re-fill once the windpump starts pumping again. Different extractable pumps employ different methods for removal of the footvalve.

In the pump illustrated in Figure 9.3 the footvalve must be fished for after the piston is removed with a special hook on a cable, and then forcibly pulled out. One common method is to utilize the pump rods to remove the footvalve. In this case the footvalve has a threaded top built into it and the piston a threaded hole under it. Once the pump rod is disconnected the piston is lowered onto the footvalve and rotated to engage the threads sections, then both piston and footvalve are removed together.

Extractable pumps are obviously preferable in deep boreholes, but the non-extractable type can be used in shallower-lift applications, where the piston diameter is, in any case, too large to allow it to be readily extracted through the rising main.

A well-maintained windpump will require fewer repairs than its poorly-maintained counterpart. Replacement parts should be easily obtainable from the manufacturer, and will be more quickly procured if that manufacturer is situated locally.

Few specialized tools are usually necessary, but it is advisable to check that none will be needed before commencing any job, particularly if the windpump is to be lowered for work to be carried out.

If major repairs are needed it is advisable to contact the manufacturer. Before doing so check if the problem is covered by the windpump warranty.

9.3 Safety

Windpumps are like other pieces of machinery: people can be injured when attending to them if safety precautions are not adhered to. It is advisable to note and use the precautions listed below to prevent this occurring.

- Read the manufacturer's safety instructions and abide by them.

- **Never** work on a windpump alone, always have at least **two people** present.

- Before climbing the machine, manually furl it and apply the brake (if one exists) unless there is good reason not to do so.

- When climbing the tower **always** use the ladder or rungs if provided, **always** wear a safety harness and **be aware** of the behaviour of the blades.

- Keep **fingers** and **toes well clear** of any moving parts.

- Try not to stand underneath the windpump if someone is working on it.

- **Do not** allow children to climb on the windpump.

- **Wear a hard hat** if possible.

To prevent damage to the machine other precautions should be taken into account:

- **Build a fence** around the windpump.

- **Protect** the delivery pipe.

- **Check** that the windpump furling mechanism works periodically.

Bibliography

General Bibliography

Deleito, J.C.C. and Cabrero, J.R., *La Energia Eolica Technologia e Historia*, Editorial Blume, Madrid, Spain.

Doorenbos, J. and Pruitt, W.O., *Crop Water Requirements*, FAO Irrigation and drainage paper 24, FAO Rome 1984. ISBN 92-5-100279-7

Eldridge, F.R., *Wind Machines*, Van Nostrand Reinhold, New York, 1980. ISBN 0-442-22194-0

ESCAP, *Wind-powered Water Pumping in Asia and the Pacific*, ST/ESCAP/1098, Economic and Social Commission for Asia and the Pacific, United Nations, New York, 1991.

Fraenkel, P.L., *Water Lifting Devices*, FAO Irrigation and drainage paper 43, FAO Rome, 1986. ISBN 92-5-102515-0

Fraenkel, P.L. and Hulscher, W.,(eds.), *The Power Guide*, IT Publications, London, 1993.

GRET/I.T.Dello, *Les Éoliennes de Pompage*, GRET, 1989. ISBN 2-86-844-030-4

Hofkes, E.H., *Renewable Energy Sources for Rural Water Supply*, technical paper 23, International Reference Centre, IRC, The Hague, The Netherlands, 1986. ISBN 90-6687-007-9

Ignacio, J. and Lus, S.U., *Energia Hidraulica y Eolica Practica*, Pamplona, Spain.

Irvin, G., *Modern Cost-Benefit Methods*, Macmillan, 1986. ISBN 0-333-23208-9

de Jong, I. (ed.), *Wind Energy for Rural Areas*, Proceedings of the international workshop, 10-14 Oct 1991, Bergen, Netherlands. Netherlands Energy Research Foundation, ECN, 1991.

Kovarik, T.,Pipher, C. and Hurst,J., *Wind Energy*, Prism Press, 1980. ISBN 0-904727-29-7

Lysen, E.H., *Introduction to Wind Energy*, Steering Committee on Wind Energy in Developing Counties (now CWD), SWD 82-1, Eindhoven University of Technology, 1982.

v. Meel, J. and Smulders, P., *Wind pumping: A Handbook*, World Bank Technical Paper 101, Industry and energy series. The World Bank, Washington DC, USA, 1989. ISBN 0-8213-1235-9

Park, J., *The Wind Power Book*, Cheshire Books, CA USA, 1981. ISBN 0-917352-06-8

Puig, J., Meseguer, C. and Cabre, M., *El Poder del Viento*, Ecotopia Ediciones, Barcelona, Spain.

Troen, I. and Petersen, L., *European Wind Atlas*, Riso National Laboratory, Denmark, 1989. ISBN 87-550-1482-8

World Meteorological Organization (WMO), *Meteorological Aspects of the Utilization of Wind as an Energy Source*, Technical note 175, WMO, Geneva, 1981.

Historical Bibliography

Baker, T.L., *A Field Guide to American Windmills*, University of Oklahoma Press, 1984. ISBN 0-8061-1901-2

Bauters, P., *Kracht van Wind en Water, Molens in Vlaanderen*, 1989. ISBN 90-6152-554-3

Eleni, Limona-Trebela, *Windmills of the Aegean Sea*, International Molinological Society, 1983.

Harverson, M., *Persian Windmills*, International Molinological Society, 1991. ISBN 90-73283-03-5

International Molinological Society, *Transactions of the International Symposium on Molinology*: No.1 Portugal 1965, No.2 Denmark 1969, No.3 Netherlands 1973, No.4 Great Britain 1977, No.5 France 1982, No.6 Belgium 1985, No.7 Germany 1989.

Jannis, C., *Windmühlen, der Stand der Forshung über das Vorkommen und denUrsprung*, Notebaart, 1972.

Rivals, C., *Le Moulin a Vent et le Meunier, dans la société francaise traditionelle*, 1987. ISBN 2-7013-0732-5

GWEP
Global Windpump Evaluation Programme

GWEP was carried out under the World Bank and the UNDP. Reports are available for individual countries, covering wind resource, history and present status of windpumping, typical applications, economic and institutional aspects, and potential for windpumps.

The country reports and a summary report are available from The Energy Department, The World Bank, 1818 H Street NW, Washington DC 20433, USA.

Countries on which GWEP reports are available are:

Africa	Asia	S. America
Botswana	China	Argentina
Cape Verde	India	Columbia
Kenya	Philippines	Peru
Morocco	Sri Lanka	
Senegal		
Tunisia		
Zimbabwe		

Glossary

Note: Cross-referenced terms are shown in italics

Acidity - A property of water (the converse of alkalinity) affecting its taste and corrosiveness. Measured on the pH scale, the acidity should usually be between pH 7 and pH 8 for drinking water.

Aerofoil - A blade or solid, curved in such a way that it produces lift when placed in an air-flow.

Air-lift pump - Introduces compressed air into the *rising main*, producing a froth of air and water which rises to the surface.

Anemometer - An instrument for measuring wind speed, usually driven by wind-induced drag forces on small revolving cups.

Aquifer - A naturally-occurring layer of water-bearing rock or sand.

Back-geared - A *transmission* that gears down the pumping speed by meshing a small gear on the *rotor* with a larger one that drives the pump.

Beaufort scale - A commonly-used measure of wind strength, defined as ranges of windspeed from force 0 to force 12.

Borehole - A narrow well (usually 4 or 6 inches in diameter) drilled into rock for the extraction of water. Used where the water level is too deep for a hand-dug well, or beneath hard rock.

Climate - The weather conditions (e.g. wind, temperature) that can be expected in a region, based on long-term averages.

Coefficient of performance, C_p - Ratio of the shaft power of the windpump to the power in the wind in the cross-sectional area of the rotor.

Crop canopy fraction - The fraction of the ground covered by the foliage of a crop.

Darrieus - A *vertical* axis *rotor* with aerofoil shaped blades, often with an 'egg-beater' or troposkein shaped profile.

Delivery pipe - The pipe from the *well-head* through which the water is discharged to the tank or distribution network.

Design month - The month whose conditions require the largest pumping system to meet them. This is the worst-case month for which the design must cater.

Direct drive - A windpump using a crank or cam such that one turn of the *rotor* relates to one whole *stroke* of the pump.

Discount factor - A factor that when multiplied by future costs or benefits, gives their *present worth*. Calculated from the *discount rate*, inflation relative to the general rate, and the number of years in the future that the payment will be made.

Discount rate - The annual rate at which the *present worth* of future costs or benefits decreases. Also known as the opportunity cost of capital. Around 10 to 12% for most economies.

Diurnal - Having a daily cycle.

Doldrums - An area near the equator in which there is little or no wind.

Drag force - Force on a body in an air-flow acting parallel to the flow.

Drawdown - The vertical distance by which the water level in a well drops once the rate of pumping is stabilized.

Drip irrigation - A field water application system using rows of perforated pipes through which water drips.

Dynamic head - The additional *head* experienced by a pump due to friction and turbulence in the pipework.

Eccentricity - The throw of the crank, or distance of the crankpin from its centre of rotation.

Economic analysis - An analysis of a project in monetary terms where costs and benefits are taken with regard to the economy as a whole. Prices exclude any local taxes, subsidies or surcharges.

Electrical transmission - System in which a wind-electric generator drives an electric pump.

Energy pattern factor - The factor that relates the actual mean power in the wind to the power that can be calculated from using the mean windspeed. It is dependent on the windspeed distribution and usually has a value of two.

Evapotranspiration - The loss of water from crops by evaporation through their leaves.

Fatigue - The process by which materials fracture when fluctuating loads are applied, even though a steady load of the same value would not cause damage.

Financial analysis - An analysis in monetary terms where costs and benefits are taken with regard to the buyer. Prices should include any local taxes and subsidies, and if applicable any costs of financing should be included.

Fishplate - Steel brackets used to join struts of windpump tower.

Footvalve - One-way valve at the base of a *piston pump* through which water is sucked on the piston up-stroke, but which closes on the down-stroke, forcing water through the piston valve.

Frequency distribution - In relation to wind measurements, the relative frequency of occurrence of different wind speeds or directions.

Furling - The mechanism (either manual or automatic) by which the windpump can be stopped or slowed, either by turning the whole *rotor* head out of the wind or by rotating each blade individually.

Geared - See *back-geared*.

Head - The height over which water must be pumped, or height of water column that is provided by a raised tank.

Header tank - A small tank to provide a constant *head* for a distribution/application system (e.g. low-head drip irrigation).

Hose-and-basin - A low-cost irrigation application system using a hand-held hose to fill earth basins around each plant.

Hot-dip galvanizing - A low-cost process providing a corrosion-resistant coating for iron or steel by 'dipping' into molten zinc.

Hub-height - The distance of the centre (i.e. the hub) of the *rotor* above ground level.

Hydraulic energy - The potential energy needed to raise a certain quantity of water through a certain *head*. Measured in Joules (J), Megajoules (MJ) or Watt-hours (Wh).

Hydraulic transmission - System which uses pumped water as a medium to transmit power from *rotor* to pump.

Impermeable - Rock through which water cannot pass.

Knots - A measure of windspeed (kn). One kn is defined as 1 nautical mile per hour (nm/h), and equal to 0.51m/s or 1.2mph.

Land breeze - A wind that can blow off the land at night in coastal areas, caused by the ground cooling the air. Tends to be weaker than the corresponding *sea breeze*.

Life-cycle cost analysis - An analysis of a project in monetary terms that takes into account all the future costs and benefits over the life-cycle of the project.

Lift force - Force on an *aerofoil* acting in a direction perpendicular to the air-flow across the aerofoil.

Low-head drip - An irrigation application system needing only about 2m *head* of water to drive it - See *drip irrigation.*

Nomogram - A diagram that can be used to perform calculations by graphical means alone.

Panamone - A *rotor* (usually *vertical axis*) which is shielded from the wind on one side and exposed on the other, operating by differential *drag.*

Payback period - The number of years necessary for the income (or value of benefits) from a project to equal the initial capital cost.

Permeability - A measure of how easily water can travel through rock or soil.

Photovoltaics (PV) - The use of solar energy to generate electricity directly via the photovoltaic effect. See *photovoltaic modules*

Photovoltaic modules - Flat panels containing semiconductor cells that are able to convert the energy in the sun's light directly into electrical energy.

Piston pump - Pump which used a piston in a cylinder to lift water - see *reciprocating pump.*

Pitman mechanism - The system used on most gear-driven windpumps to convert the rotary motion of the gears into a reciprocating motion at the *pump-rod*. Consists of a connecting rod located on a gear wheel at one end, and at the other hinging on a guided vertical slider attached to the pump-rod.

Pneumatic transmission - System in which the windmill produces compressed air which is then used to drive a pump.

Present worth - The value of a future cost or benefit in current terms, after being adjusted for future changes in the value of money.

Prevailing wind - The normal wind direction for a region.

Progressive cavity pump - A rotary pump using a solid helical rotor in a flexible stator.

Pump rod - The rods that link the pump down the well to the *rotor* and *transmission* assembly at the top of the tower. The action of the pump rod is a vertical reciprocating motion.

PV pumping - Water pumping using solar *photovoltaics* as a power source for an electric pump.

Rate of return - The annual income (or value of benefits) from a project expressed as a percentage of initial investment costs.

Rated windspeed - Design maximum windspeed at which the pump should perform optimally.

Reciprocating pump - A pump that is operated by a 'forwards and backwards' motion. These are usually piston pumps and are almost universally used with windpumps due to their high efficiency at high *heads*.

Rising main - The tube through which the water is lifted to the surface after leaving the pump. The *pump-rods* also usually pass down this tube, entering through a seal at the *well head.*

Rolling-element bearings - Bearings containing a rolling component (e.g. ball or needle bearings) as opposed to plain bearings or bushes.

Rotor - The assembly consisting of the windpump blades and their supports that convert the wind's motion to rotary shaft power.

Rotary pump - A pump driven by a rotating shaft, such as a centrifugal pump or a screw pump. Very rarely used with windpumps.

Run-of-wind - Used as a means of measuring windspeed, the run-of-wind is the distance an air parcel would travel in a certain time.

Salinity - A measure of the salt content of water, affecting both taste and corrosiveness.

Savonius rotor - 'S' shaped *rotor* consisting of two or three curved blades interlocking around a central shaft. Works on a mixture of *drag forces* and *lift forces.*

Sea breeze - A wind which blows off the sea during the daytime (particularly during the afternoon) in coastal areas. Caused by the heating and rising of the air over the adjacent land.

Solar pumping - See *PV pumping*.

Solidity - Ratio of total of the blade widths to the circumference of the *rotor*.

Specific yield - A measurement used to compare the output of different *boreholes*, calculated as the pumping rate divided by the *drawdown* at that rate.

Static water level - The water level in a well or *borehole* before pumping begins.

Stroke - The maximum extent of travel of the *pump rods* and piston.

Subsoil - The layer of weathered material that underlies the surface soil.

Sucker rod - Another name for a *pump-rod*.

Surface roughness - A measure of ability of the surface to slow the wind at low levels by turbulence and friction, and so affect the profile of the windspeed with height.

Surface type - A rough way of classifying different terrains according to their *surface roughness*.

Tip-speed ratio - Ratio of the blade tip speed to the windspeed. Higher for lower *rotor solidity*.

Topography - The shape of the landscape (i.e. hills, valleys, ridges).

Torque - Turning force or moment

Transmission - The mechanism by which the rotary motion of the *rotor* is converted to a reciprocating motion at the *pump-rod*, often involving a gearing-down of the *rotor* speed.

Tubewell - Alternative name for a borehole.

Turbulence - The random, gusty motions set up in an air-stream near obstacles such as trees or buildings. Has an adverse effect on windpump efficiency. Also causes energy losses in water pipes, adding to the *dynamic head*.

Vertical-axis windpump - A machine in which the blades rotate on a vertical shaft rather than a horizontal one. Very rarely used with windpumps.

Volumetric efficiency - The efficiency of water distribution, expressed as percentage of water pumped that reaches the point of use (e.g. a volumtric efficiency of 90% implies that 10% of the water pumped is lost).

Water table - The level below the ground at which the natural water level can be found. The same as the *static water level*.

Weibull distribution - A mathematical means of describing the various possible statistical windspeed distributions at different sites.

Weibull parameters - The coefficients used in the Weibull equations that define the shape of the windspeed distribution.

Well casing - The outer casing of the well or *borehole* that prevents in-fill of sand or collapse of the well walls.

Well head - The point at the surface of the well where the *pump-rods* enter the *rising main* usually through a seal (sometimes called a stuffing-box), and the *delivery pipe* carries the water away.

Wind vane - A simple pointing device to indicate the direction of the wind.

Wind-wheel - Alternative name for the *rotor* of a windpump.

Wind turbine - Term generally used to describe a wind-powered electric generator.

Yaw tube - Vertical tube about the which the *rotor/transmission/tail* assembly rotates and through which pass the *pump rods*.

Yawing - The motion of the *rotor* head about its vertical pivot, e.g. to point into the wind.

APPENDIX A

Manufacturers and suppliers of windpumps by continent

AFRICA

Bobs Harries Engineering
Karamaini Estate
PO Box 40
Thika
Kenya
tel +254 151 47234/47250
fax +254 2 332009
telex 23161 KIJITO
telegrams: Bobs-Thika

Pwani Fabricators
PO Box 88734
Mwabundu Rd., Industrial Area
Mombasa
Kenya
tel +254 11 495520/493408
fax +254 11 493406

Siden
6 rue Ibn Inane
Rabat
Morocco

Climax Windmills
PO Box 20244
Peacehaven 1934
South Africa
tel +27 16 42351
fax +27 16 43141
telex 743106

Sahara Engineering Co.
PO Box 2220
Khartoum
Sudan
telex 22980 Attia

Sinaes International
2ème Porte Mermoz
Dakar
Senegal
tel +221 325 280

Louis Guillaud
31 rue Pierre Parent
Casablanca
Morocco
tel +212 2 305971/307570

Serept Energies Nouvelles
66 Route de l'Ariana 4km
2080 Ariana
Tunisia
tel +216 1 750 164
fax +216 2 56 183
telex 13042 TN

**South Africa Plant &
Engineering Co.**
PO Box 6159
Dunswart 1508 Transvaal
South Africa
tel +27 52 4327/4328/4377
fax +27 52 5400
telex 83119

ISERST
BP 486
Djibouti
tel +253 35 27 95
fax 253 35 42 92
telex 5811 DJ

Sheet Metal Kraft
14 Coventry Street
L.I.S. Belmont
PO Box 1840, Bulawayo
Zimbabwe
tel +263 9 74100/74106
fax +263 9 74735

Stewarts & Lloyds
37 Leopold Takawira St.
PO Box 784, Harare
Zimbabwe
tel +263 0 708 191
fax +263 0 790 972
telex 24261 ZW

Anant Energy Systems Ltd
4/5 Neptune Tower
Opp. Nehrubridge, Ashram Rd
Ahmedabad 380 009
Gujurat
India
tel +91 272 407295/556
fax +91 272 405027

Aureka
Auroville 605 101
Tamilnadu
India
tel +91 413 862278
fax +91 413 862185
telex 496272 ADPS IN

Auto Spares Industries
C-7 Industrial Estate
Pondicherry
605 001
India
tel +91 413 28791/23174

Bharat Heavy Electricals Ltd
Corporate R&D Division
Vikas Nagar
Hyderabad 500 593
India
tel +0842-261988
fax +0842-263320
telex 4256404

Gram Swarajya Sikshan Kendra
Gopaldham
PO Ghela Somnath
Ta. Jasdan
Dist. Rajkot
Gujurat 360 050
India

MERIN (PVt.) Ltd.
Agrotool Division, LA/8a
Block 22, Federal B Area
Karachi
Pakistan
tel +92 21 681517/674639
fax +92 21 2417836
telex 24675 MERIN PK

Nei Menggu Shangdu Graziery Machinery Factory
Shangdu 013450
China
tel +86 28 31 231

NEPC-MICON Ltd
No. 3 Goomes St.
Madras 600 001
India
tel +91 44 587403/581642
fax +91 44 586112
telex 417607

Reymill Steel Products
Sta Rosa
Nueva Ecija
The Philippines
tel +63 662 518

Star Engineers
561 Jaffna Junction
Anuradhapura
Sri Lanka
tel +94 25 2571

USA Econ. Devel. Co. Ltd.
56/7 Prachachuen Road
Bankhen District
Bangkok 10210
Thailand
tel +66 2 5890935/5892221
fax +66 2 5890935

Water Resources Board
PO Box 34
2A Gregory's Avenue
Colombo 7
Sri Lanka
tel +94 1 697050/694835
telegrams JALASAMPAT

Wind Fab (Mayee Eng. Ltd.)
447 Avanashi Road
Peelamedu
Coimbatore 641 004
India
tel +91 422 36079

Ferguson Manufacturing
64 Main Road
Kumeu
New Zealand
tel +64 9 412 8655
fax +64 9 412 7037

W.D. Moore Ltd.
3 Keegan Street
O'Connor, 6163
Western Australia
tel +61 9 337 4766
fax +61 9 314 1306

Wire Makers Limited
120 Maces Road
PO Box 244
Christchurch
New Zealand
tel +64 3 3842 069
fax +61 3 3842 569

Jolly Windmill Company
PO Box 15-560
New Lynn
Auckland 7
New Zealand
tel +64 9 818 8696

Southern Cross Machinery
PO Box 155
Darra
Queensland 4076
Australia
tel +61 7 375 3944
fax +61 7 375 4553
telex 41033

AbaChem Engineering Ltd.
Northern Road
Newark, Notts.
NG24 2EH
UK
tel +44 636 76483
fax +44 636 708632

Energomachexport
25A Bezbozhny, per.
129010, Moscow
Russia
tel +7 095 288 8456
fax +7 095 288 7990
telex 411965

Mid Wales Productions Ltd.
Westgate Street
Llanidloes, Powys
SY18 6HN
UK
tel +44 551 22104
fax +44 551 23673

BJ - Steel
Kaerbyvej 3
8832 Skals
Denmark
tel +45 86 69 44 77
fax +45 86 69 50 69

D. Hermeneau
32 Rue Marcellin-Berthelot
Z.I. des Touches
53000 Laval
France
tel +33 43 53 65 90

Molinos Company Ltd.
46 Gertsena St.
PO Box 146
Moscow
Russia
tel +7 095 1580249/1584409
fax +7 095 1582977
telex 411021 AGROS

J. Bornay Aerogeneradores
Avda. de Ibi, 76-78
03420 Castalla
Spain
tel +34 6 556 0025
fax +34 6 556 0752

LMW Windenergy B.V.
Lijnbaanstraat 1A
9711 RT, Groningen
The Netherlands
tel +31 50 145229
fax +31 50 146293
telex 20010 PMS NL

Neue Energien Wiehengebirge
Mindener Str 205
W-4500 Osnabrück
Germany
tel +49 541 7102 175
fax +49 541 7102 176

NM-Electro
Fruevej 80
DK-7900
Nykobing Mors
Denmark
tel +45 9772 5200
fax +45 9772 5200

Poncelet et Cie
Place de la Victoire
10380 Plancy l'Abbaye
BP 12
France
tel +33 25 37 40 15
fax +33 25 37 43 72

Power Mills
Toftegaardsvej 32
DK-9900
Frederikshavn
Denmark
tel +45 9843 0200
fax +45 9843 8177

Servicios Electronicos Sola
C/. Coll i Vehi, 49-51
17100 La Bisbal (Girona)
Spain
tel/fax +34 72 640894

Tozzi & Bardi
Via Norvegia 16
58100
Grosseto
Italy
tel +39 564 28401

Vergnet
66 rue Hoche
92240 Malakoff
France
tel +33 1 47 46 16 16
fax +33 1 47 46 06 86
telex 632295 VERGNET F

NORTH AMERICA

Aermotor Windmill Corp.
3015 North Bryant
Box 5110
San Angelo, TX
76902, USA
tel +1 915 658 2795
fax +1 915 655 7147

Bergey Windpower Co.
2001 Priestly Ave
Norman, OK
73069
USA
tel +1 405 364 4212
fax +1 405 364 2078
telex 6502987196 MCI UW

Dempster Industries Inc.
Box 848, 711 S. 6th St.
Beatrice, NE
68310
USA
tel +1 402 223 4026
fax +1 402 228 4389
telex 701447

Essex Associates Inc.
PO Box 540441
Dallas, TX
75220-2411
USA
tel +1 214 556 1317
fax +1 214 869 9051

KMP Pump Co.
Box 220
Earth, TX
79031
USA
tel +1 806 257 3411

Northern Power systems
One North Wind Road
Moretown, VT
05660-0659
USA
tel +1 802 496 2955
fax +1 802 496 2953

Thermax Corporation
PO Box 3128
Burlington, VT
05401-3128
USA
tel +1 802 658 1098
fax +1 802 658 1098

Máquinas Agrícolas Alegranense Ltda
Praca Oswaldo Aranna 160
Caixa Postal 22
97 540 000 Alegrete RS
Brazil
tel +55 422 2412/2925
fax +55 422 2412

Máquinas Agrícolas de Agua
Rua Visconde de Ouro, 865
Parque Industrial
15 030 000
Sao Jose do Rio Preto SP
Brazil
tel +55 172 322 511

Empresa Mecánica de Camaguey
Misisterio de Sidero Mecánica
c/o CEN
Calle E no.261
Habana
Cuba

Juan Cardoza
Jirón Los Andes Lote 3
Castilla
Piura
Peru

Cataventos Kenya Ltd.
Rodovia RS 130
Km 14, PO Box 111
Encantado-RS
Brazil
tel +55 51 751 1750
fax +55 51 751 1471
telex 510115 KNYA

Facogsa S.R.L.
Jr. Horacio Patino 196
Urb Latino
Chiclayo
Peru
tel +51 74 242682
fax +51 74 231989

FIASA
Horoigeura 1890
Buenos Aries
Argentina
tel +54 1 923 1081/5
fax +54 1 924 1648
telex 9028 FIASA AR

Centro Las Gaviotas
Paseo Bolivar No 20-90
Avda Circunvalar
Bogota
Columbia
tel +57 1 286 2876
fax +57 1 268 4742

Hidromec
Cota Calle 30, No 100
La Paz
PO Box 7428
Bolivia
tel +591 2 795725/797958

Industrias Metalicas Indusierra
Carrera 64 A No 6-12
Bogota
PO Box 080990
Columbia
tel +57 1 260 3043

Industrias Metalurgicas del Pueblo (IMEP)
Km. 5-1/2 Carretera norte
300 vrs. al lago, A.P. 2409
Managua
Nicaragua
tel +505 2 40906/40103
fax +5050 2 40906

Juan Jesus Avila Nole
Sector Maestranza
Miramar
Piura
Peru

Industrias Jober Ltda
Calle 20 No.30-104
Duitama
Columbia
tel +57 987 603887

Molinos de Viento
Pachacutec
Urb. Semi Rural Pachacutec
Av Circunvalación y Moquegua
Arequipa
Peru
tel +51 54 222 617

Lázaro Vargas
800 Norte 150 Este
Palacio de los Deportes
Heredia
Costa Rica

Oficina Técnica
Arnaldo Maudet
Calle Real de Sabana Grande
Edificio Celeste
Piso 9, Oficina 28
Caracas 1050
Venezuela
tel +58 2 723343/716127
fax +58 2 720772

Ferreteria Santiago sa
Av Bernardo O'Higgins 1796
Santiago
Chile
tel +56 2 671 2077
fax +56 2 696 6213

Ricardo Zimic Vidal
Av Republica de Panama 5252
Lima 34
Peru
tel +51 14 466 395

APPENDIX B

Buyers' guide to products

During 1991/1992 all known windpump manufacturers worldwide were contacted and asked to supply information concerning their products and services. Their replies are summarized in this appendix. The entries are grouped into continents and within this listed alphabetically. Photographs have been included where they were supplied by the manufacturer.

The data in each entry are fairly self explanatory; however, 'Drive type' is described by the letters G, D or E, defined as follows:

G   Gear-driven
D   Direct drive (i.e. crank)
E   Electric (i.e. wind generator and
           electric pumpset)

Costs have all been shown in US dollars for consistency. Where cost data was supplied in local currency this has been converted at 1992 exchange rates. Manufacturers were asked to supply a price for a complete system, including wind-head assembly, standard height tower and pump. In reality the items covered by the given price may differ slightly between manufacturers, and so the prices should only be taken as approximate.

Rotor diameters and maximum heads have been converted to metres where necessary.

It should be realized that the completeness and accuracy of the data and the quality of photographs can only be as good as has been supplied by the manufacturers.

Bobs Harries Engineering

Karamaini Estate
PO Box 40
Thika
Kenya

tel +254 151 47234/47250
fax +254 2 332009
telex 23161 KIJITO
telegrams: Bobs-Thika

The Kijito was developed
from the I T Windpump

TRADE NAME	Kijito				
Rotor diameter (m)	2	3.7	4.9	6.1	7.4
Number of blades	6	18	24	24	24
Drive type	D	D	D	D	D
Max pumping head (m)	24	200	240	240	240
Cost ($US)	-	-	-	-	-

Climax Windmills

PO Box 20244
Peacehaven 1934
South Africa

tel +27 16 42351
fax +27 16 43141
telex 743106

Type 'R' pumps use a high
speed rotary drive

TRADE NAME : Climax	No8	No10	No12	No14	No18	15R	18R
Rotor diameter (m)	2.44	3.05	3.66	4.27	5.49	4.57	5.49
Number of blades	18	18	18	18	24	18	24
Drive type	G	G	G	G	D	R	R
Max pumping head (m)	150	150	150	150	150	75	75
Cost ($US)	1848	1888	2540	2711	5104	4348	5356

Institut Superieur d'Etudes et des Recherches Scientifiques et Techniques

BP 486
Djibouti

tel +253 35 27 95
fax +253 35 42 92
telex 5811 DJ

TRADE NAME	CWD 2740
Rotor diameter (m)	2.7
Number of blades	6
Drive type	D
Max pumping head (m)	6
Cost ($US)	2327

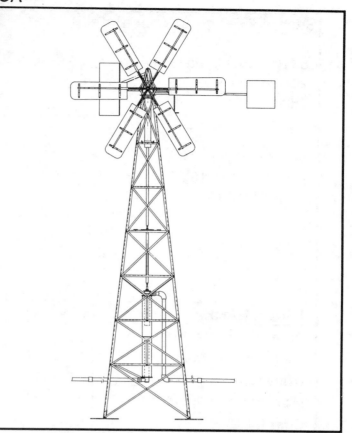

Pwani Fabricators

PO Box 88734
Mwabundu Road
Industrial Area, Mombassa
Kenya

tel +254 11 495520/493408
fax +254 11 493406

TRADE NAME	Pwani	
Rotor diameter (m)	3.6	4.9
Number of blades	18	18
Drive type	G	G
Max pumping head (m)	-	-
Cost ($US)	-	-

Sahara Engineering Company

PO Box 2220
Khartoum
Sudan

telex 22980 Attia

TRADE NAME	CWD 5000
Rotor diameter (m)	5
Number of blades	8
Drive type	D
Max pumping head (m)	100
Cost ($US)	

Serept Energies Nouvelles

66 Route de l'Ariana 4km
2080 Ariana
Tunisia

tel +216 1 750 164
fax +216 2 56 183
telex 13042 TN

TRADE NAME	SEN 5000
Rotor diameter (m)	5
Number of blades	8
Drive type	D
Max pumping head (m)	50
Cost ($US)	5000

Sheet Metal Kraft

14 Coventry Street
L.I.S. Belmont
PO Box 1840, Bulawayo
Zimbabwe

tel +263 9 74100/74106
fax +263 9 74735

TRADE NAME	M.K.
Rotor diameter (m)	3.6
Number of blades	18
Drive type	G
Max pumping head (m)	104
Cost ($US)	2800

South Africa Plant & Engineering Co.

PO Box 6159
Dunswart 1508 Transvaal
South Africa

tel +27 52 4327/4328/4377
fax +27 52 5400
telex 83119

The Panemone is a vertical-axis turbine
driving a rotary positive displacement pump

TRADE NAME	Panemone 1000
Rotor diameter (m)	2.3
Number of blades	3 tier open "s"
Drive type	belt
Max pumping head (m)	250
Cost ($US)	-

Stewarts and Lloyds

37 Leopold Takawira St.
PO Box 784, Harare
Zimbabwe

tel +263 0 708191
fax +263 0 790972
telex 24261 ZW

TRADE NAME	IT Windpump
Rotor diameter (m)	6
Number of blades	24
Drive type	D
Max pumping head (m)	200
Cost ($US)	3800

Aureka

Auroville 605 101
Tamilnadu
India

tel +91 413 862278
fax +91 413 862185
telex 496272 ADPS IN

TRADE NAME	AV 55
Rotor diameter (m)	5.5
Number of blades	24
Drive type	D
Max pumping head (m)	100
Cost ($US)	1751

Auto Spares Industries

C-7 Industrial Estate
Pondicherry
605 001
India

tel +91 413 28791/23174

TRADE NAME	Wind Machines			
Rotor diameter (m)	2	3	4	5
Number of blades	12	12	12	12
Drive type	G	G	G	G
Max pumping head (m)	30	75	110	300
Cost ($US)	1362	1751	4280	6226

Bharat Heavy Electricals Ltd

Corporate R&D Division
Vikas Nagar
Hyderabad 500 593
India

tel +91 842 261988
fax +91 842 263320
telex 4256404

TRADE NAME	BHEL
Rotor diameter (m)	5
Number of blades	12
Drive type	D
Max pumping head (m)	50
Cost ($US)	778

Merin (PVt.) Ltd.

Agrotool Division, LA/8a
Block 22, Federal B Area
Karachi
Pakistan

tel +92 21 681517/674639
fax +92 21 2417836
telex 24671 MERIN PK

The Tawana was developed
from the I T Windpump

TRADE NAME	Tawana		Zorawar
Rotor diameter (m)	6	7	3.6
Number of blades	24	24	18
Drive type	D	D	G
Max pumping head (m)	200	200	30
Cost ($US)	10204	10204	3878

NEPC-MICON Ltd

No. 3 Goomes St.
Madras
600 001
India

tel +91 44 587403/581642
fax +91 44 586112
telex 417607

TRADE NAME	Green Rev
Rotor diameter (m)	3
Number of blades	18
Drive type	G
Max pumping head (m)	84
Cost ($US)	1829

Star Engineers

561 Jaffna Junction
Anuradhapura
Sri Lanka

tel +94 25 2571

TRADE NAME	NIVA 3000
Rotor diameter (m)	3
Number of blades	6
Drive type	D
Max pumping head (m)	10
Cost ($US)	612

U.SA Economic Development Co. Ltd.

56/7 Prachachuen Road
Bankhen District
Bangkok 10210
Thailand

tel +66 2 5890935/5892221
fax +66 2 5890935

TRADE NAME	USA windmill						
Rotor diameter (m)	2.44	3.05	3.66	4.25	4.8	5.5	6
Number of blades	30	30	30	30	40	40	40
Drive type	-	-	-	-	-	-	-
Max pumping head (m)	-	-	-	-	-	-	-
Cost ($US)	-	-	-	-	-	-	-

Wind Fab (Mayee Eng. Ltd.)

447 Avanashi Road
Peelamedu
Coimbatore 641 004
India

tel +91 422 36079

TRADE NAME	Blue Vane
Rotor diameter (m)	3.05
Number of blades	18
Drive type	G
Max pumping head (m)	60
Cost ($US)	1556

Southern Cross Machinery Pty Ltd.

PO Box 155
Darra, Qld
4076
Australia

tel +61 7 375 3944
fax +61 7 375 4553
telex 41033

TRADE NAME	IZ Pattern windmills					R Pattern windmills		
Rotor diameter (m)	1.8	2.4	3	3.6	4.2	5.1	6.3	7.5
Number of blades	10	18	18	18	18	24	30	36
Drive type	G	G	G	G	G	D	D	D
Max pumping head (m)	22	40	72	96	135	146	175	216
Cost ($US)	1385	1430	2285	2675	2775	7070	10605	14046

Wire Makers Limited

120 Maces Road
PO Box 244
Christchurch
New Zealand

tel +64 3 3842 069
fax+64 3 3842 569

TRADE NAME	Hayes
Rotor diameter (m)	2
Number of blades	8
Drive type	D
Max pumping head (m)	80
Cost ($US)	1333

W.D. Moore Ltd.

3 Keegan Street
O'Connor, 6163
Western Australia

tel+61 9 337 4766
fax+61 9 314 1306

TRADE NAME	Yellowtail					Aermotor			
Rotor diameter (m)	1.8	2.4	3	3.6	4.2	1.8	2.4	3	3.6
Number of blades	18	18	18	18	24	18	18	18	18
Drive type	G	G	G	G	G	G	G	G	G
Max pumping head (m)	22	39	81	89	117	29	43	66	98
Cost ($US)	1906	2268	2929	3670	5151	2157	2738	3686	5304

EUROPE

AbaChem Engineering

Northern Road
Newark, Notts.
NG24 2EH
UK

tel+44 636 76483
fax+44 636 708632

TRADE NAME	AbaChem Windpump							
Rotor diameter (m)	2	2.5	3	3.75	4.25	5	5.5	6.25
Number of blades	15	15	18	21	21	21	24	24
Drive type	G	G	G	G	G	G	G	G
Max pumping head (m)	7.5	30	45	75	100	120	120	120
Cost ($US)	6470	6547	6758	7796	8477	8915	9354	9913

BJ - Steel

Kaerbyvej 3
8832 Skals
Denmark

tel+45 86 69 44 77
fax+45 86 69 50 69

TRADE NAME	BJ360	LJ3600
Rotor diameter (m)	3.6	3.6
Number of blades	8	8
Drive type	D	E
Max pumping head (m)	8	120
Cost ($US)	5424	6102

J Bornay

Avda de Ibi, 76-78
03420 Castalla
(Alicante)
Spain

tel +34 6 556 0025
fax +34 6 556 0752

TRADE NAME	Aeromotor					
Rotor diameter (m)	1.8	2.4	3.0	3.6	4.2	4.8
Number of blades	18	18	18	18	18	18
Drive type	G	G	G	G	G	G
Max pumping head (m)	23	34	52	77	110	180
Cost ($US)	-	-	-	-	-	-

Energomachexport

25A Bezbozhny, per.
129010, Moscow
Russia

tel+7 095 288 8456
fax+7 095 288 7990
telex 411965

AMB1.2 uses diaphagm pump
YBM2/3 use piston pump
YBM4/YB3B6 use screw pump

TRADE NAME	AMB1.2	YBM-2	YBM-3	YBM-4	YB3B6
Rotor diameter (m)	1.2	2	3	4	6.6
Number of blades	12	24	4	2	2
Drive type	D	D	-	G	E
Nominal pumping head (m)	10	20	20	30	30
Cost ($US)	475	-	3100	-	-

Ets. D. Hermeneau

32 Rue Marcellin-Berthelot
Z.I. des Touches
53000 Laval
France

tel+33 43 53 65 90

TRADE NAME	Mistral		
Rotor diameter (m)	2	2.25	2.5
Number of blades	15	15	15
Drive type	D	D	D
Max pumping head (m)	20	30	30
Cost ($US)	2190	2325	2960

LMW Windenergy B.V.

Lijnbaanstraat 1A
9711 RT, Groningen
The Netherlands

tel+31 50 145229
fax+31 50 146293
telex 20010 PMS NL

TRADE NAME	1003	2500
Rotor diameter (m)	3	5
Number of blades	3	3
Drive type	E	E
Max pumping head (m)	120	120
Cost ($US)	9412	14071

Mid Wales Productions

Westgate Street
Llanidloes, Powys
SY18 6HN
UK

tel+44 551 22104
fax+44 551 23673

TRADE NAME	MWW					
Rotor diameter (m)	1.8	2.4	3	3.6	4.2	4.8
Number of blades	18	18	18	18	18	24
Drive type	G	G	G	G	G	G
Max pumping head (m)	30	37.5	60	105	112.5	120
Cost ($US)	-	-	-	-	-	-

NM-Electro

Fruevej 80
DK-7900
Nykobing Mors
Denmark

tel/fax +45 9772 5200

Developed in conjunction with the
Danish Folkecenter

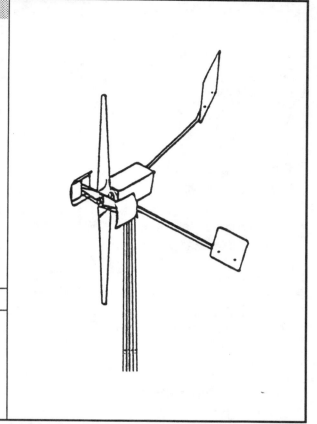

TRADE NAME	FC 4000
Rotor diameter (m)	4
Number of blades	2
Drive type	E
Max pumping head (m)	-
Cost ($US)	-

Poncelet et Cie

Place de la Victoire
10380 Plancy l'Abbaye
BP 12
France

tel+33 25 37 40 15
fax+33 25 37 43 72

TRADE NAME	Oasis			
Rotor diameter (m)	1.75	2	2.25	2.5
Number of blades	15	15	15	15
Drive type	D	D	D	D
Max pumping head (m)	10	20	30	40
Cost ($US)	1187	1278	1433	1429

| **Power Mills**

Toftegaardsjej 32
DK-9900
Frederikshavn
Denmark

tel +45 9843 0200
fax +45 9843 8177

Manufacture electrical (10 kW)
and mechanical windpumps

TRADE NAME	Power Mills	
Rotor diameter (m)	6.30	5.30
Number of blades	16	16
Drive type	E	D
Max pumping head (m)	-	-
Cost ($US)	18745	8507

Servicios Electronicos Sola (SES)

C/. Coll i Vehi, 49-51
17100 La Bisbal (Girona)
Spain

tel/fax +34 72 640894

SES manufacture a range
of windpumps

TRADE NAME	SES
Rotor diameter (m)	-
Number of blades	16
Drive type	-
Max pumping head (m)	180
Cost ($US)	-

EUROPE

Vergnet SA

66 rue Hoche
92240 Malakoff
France

tel+33 1 47 46 16 16
fax+33 1 47 46 06 86
telex 632295 VERGNET F

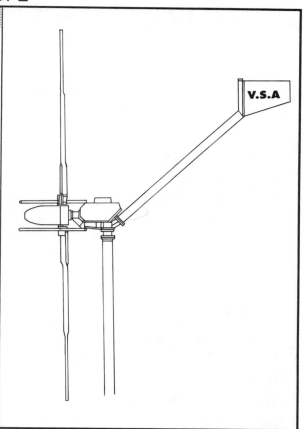

TRADE NAME	Vergnet Sa.
Rotor diameter (m)	10
Number of blades	2
Drive type	E
Max pumping head (m)	300
Cost ($US)	37644

NORTH AMERICA

Aermotor Windmill Corporation

3015 North Bryant
Box 5110
San Angelo, TX
76902, USA

tel+1 915 658 2795
fax+1 915 655 7147

TRADE NAME	Aermotor					
Rotor diameter (m)	1.8	2.4	3	3.6	4.2	4.8
Number of blades	18	18	18	18	18	18
Drive type	G	G	G	G	G	G
Max pumping head (m)	40	55	85	130	180	300
Cost ($US)	3290	3590	4740	6770	8800	11980

Bergey Windpower Co.

2001 Priestly ave
Norman, OK
73069
USA

tel +1 405 364 4212
fax +1 405 364 2078
telex 6502987196 MCI UW

TRADE NAME	BWC 1500-PD	BWC EXCEL-PD
Rotor diameter (m)	3.05	7
Number of blades	3	3
Drive type	E	E
Max pumping head (m)	130	225
Cost ($US)	5030	17500

Dempster Industries Inc.

Box 848, 711 S. 6th St.
Beatrice, NE
68310
USA

tel +1 402 223 4026
fax +1 402 228 4389
telex 701447

TRADE NAME	Dempster	
Rotor diameter (m)	1.8	4.2
Number of blades	15	24
Drive type	G	G
Max pumping head (m)	216	216
Cost ($US)	4100	4100

Essex Associates Inc.

PO Box 540441
Dallas, TX
75220-2411
USA

tel+1 214 556 1317
fax+1 214 869 9051

TRADE NAME	Fiasa					
Rotor diameter (m)	1.8	2.4	3	3.6	4.2	4.8
Number of blades	18	18	18	18	18	18
Drive type	G	G	G	G	G	G
Max pumping head (m)	52	74	112	168	240	400
Cost ($US)	3030	3375	4440	6995	9312	12624

KMP Pump Co.

Box 220
Earth, TX
79031
USA

tel+1 806 257 3411

TRADE NAME	Parish Windmill	
Rotor diameter (m)	2.4	3
Number of blades	18	18
Drive type	D	D
Max pumping head (m)	73	108
Cost ($US)	2488	3120

NORTH AMERICA

Thermax Corporation

PO Box 3128
Burlington, VT
05401-3128
USA

tel+1 802 658 1098
fax+1 802 658 1098

The Helius Water Lifter is a
vertical-axis machine

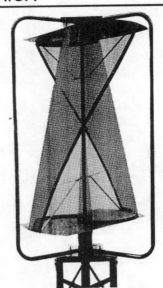

TRADE NAME	Helius Water Lifter				Windstream Electric
Rotor diameter (m)	0.6	1	1.5	2	1
Number of blades	2	2	2	2	2
Drive type	G	G	G	G	E
Max pumping head (m)	30	30	30	30	30
Cost ($US)	480	1200	3000	4800	590

SOUTH AMERICA

Cataventos Kenya Ltd.

Rodovia RS 130
Km 14, PO Box 111
Encantado-RS
Brazil

tel+55 51 751 1750
fax+55 51 751 1471
telex 510115 KNYA

TRADE NAME	Kenya	Portable Kenya Junior
Rotor diameter (m)	3	1.6
Number of blades	18	6
Drive type	G	G
Max pumping head (m)	80	15
Cost ($US)	980	-

FIASA- FAB

Horoigeura 1890
Buenos Aries
Argentina

tel+54 1 923 1081/5
fax+54 1 924-1648
telex 9028 FIASA AR

TRADE NAME	FIASA Windmill					
Rotor diameter (m)	1.8	2.4	3	3.6	4.2	4.8
Number of blades	18	18	18	18	18	18
Drive type	G	G	G	G	G	G
Max pumping head (m)	39	56	84	126	180	300
Cost ($US)	1150	1315	1735	2610	3900	5055

Industrias Jober LtD.

Calle 20 No. 30-104
Duitama
Columbia

tel+57 987 603887

TRADE NAME	Jober		Hollman JBR-3000
Rotor diameter (m)	2.5	2.5	3
Number of blades	10	16	18
Drive type	D	D	G
Max pumping head (m)	-	-	-
Cost ($US)	8100	9500	-

Facogsa S.R.L.

Jr. Horacio Patino 196
Urb Latino
Chiclayo
Peru

tel+51 74 242682
fax+51 74 231989

TRADE NAME	Alborada
Rotor diameter (m)	4.5
Number of blades	8
Drive type	D
Max pumping head (m)	-
Cost ($US)	1700

During the 1991/92 survey, all manufacturers were asked to provide information on other related products and services that they offered.

The replies from those that answered that section of the questionnaire are summarized in the table on the following page. A box indicates that the service is provided. Manufacturers that did not complete this section have been omitted from the list.

Manufacturers were questioned regarding the six areas listed opposite, which are abbrieviated in the column headings of the table.

- Site evaluation

- System selection

- System installation

- Regular maintenance

- Manufactures pumps

- Will supply stand-alone wind-electric generators

- Will supply grid-connected wind-electric generators

COMPANY	SERVICES OFFERED						
	Site eval	System select	System inst	Regular maint	Mftr pumps	S/alone gens	Grid con gens

See page C1 for full explanation

AFRICA

Company	Site eval	System select	System inst	Regular maint	Mftr pumps	S/alone gens	Grid con gens
Climax Windmills	❑	❑					
ISERST	❑	❑	❑	❑	❑		
Sahara Engineering Co.			❑	❑	❑		
Serept Energies Nouvelles	❑	❑	❑	❑	❑		
Sheet Metal Kraft	❑	❑	❑	❑	❑		
Stewarts and Lloyds	❑	❑	❑	❑			

ASIA

Company	Site eval	System select	System inst	Regular maint	Mftr pumps	S/alone gens	Grid con gens
Aureka	❑	❑	❑	❑	❑		
Auto Spare Industries	❑	❑	❑	❑	❑		
Baharat Heavy Electricals Ltd	❑	❑	❑			❑	❑
MERIN (PVt.) Ltd.	❑	❑	❑	❑	❑	❑	
NEPC-MICON Ltd	❑	❑	❑	❑		❑	❑
Star Engineers		❑	❑	❑	❑		
Water Resources Board	❑	❑	❑	❑	❑		
Wind Fab (Mayee Eng. Ltd.)	❑	❑	❑	❑	❑		

AUSTRALIA

Company	Site eval	System select	System inst	Regular maint	Mftr pumps	S/alone gens	Grid con gens
W.D. Moore Ltd.	❑	❑	❑		❑	❑	❑
Southern cross pty	❑	❑	❑		❑		
Wire Makers Ltd.		❑		❑			

EUROPE

Company	Site eval	System select	System inst	Regular maint	Mftr pumps	S/alone gens	Grid con gens
AbaChem Engineering Ltd.	❑	❑	❑	❑	❑		
BJ - Steel	❑	❑	❑	❑	❑	❑	❑
Energomachexport						❑	❑
LMW Windenergy B.V.	❑	❑	❑			❑	❑
Mid Wales Productions	❑	❑	❑	❑	❑		
Power Mills	❑	❑	❑	❑	❑	❑	
Vergnet Sa.	❑	❑	❑	❑		❑	❑

N AMERICA

Company	Site eval	System select	System inst	Regular maint	Mftr pumps	S/alone gens	Grid con gens
Bergey Windpower Co.	❑	❑	❑	❑		❑	❑
Dempster Industries Inc.		❑			❑		
Essex Associates Inc.		❑			❑		
KMP Pump Co.	❑	❑	❑	❑		❑	❑
Thermax Corporation	❑	❑	❑	❑		❑	

S AMERICA

Company	Site eval	System select	System inst	Regular maint	Mftr pumps	S/alone gens	Grid con gens
Cataventos Kenya Ltd.	❑	❑	❑	❑	❑	❑	
FIASA FAB	❑	❑	❑	❑			
Industrias Jober Ltda.	❑	❑	❑	❑	❑		

APPENDIX D

Discount factors for life-cycle cost analysis

Single future costs or benefits

Table D.1 is used to find the discount factor for a single cost or benefit at some point in the future. The future cost should be multiplied by the appropriate factor to give its present worth. A description and examples of the use of this and the following table for the calculation of present worth is given in section 7.2.

Table D.1 Discount factors for a single future payment

Discount rate (d)	Inflation rate (i)	Factor Pr for given number of years				
		5	10	15	20	30
	0.00	1.00	1.00	1.00	1.00	1.00
	0.02	1.10	1.22	1.35	1.49	1.81
0.00	0.04	1.22	1.48	1.80	2.19	3.24
	0.06	1.34	1.79	2.40	3.21	5.74
	0.08	1.47	2.16	3.17	4.66	10.06
	0.10	1.61	2.59	4.18	6.73	17.45
	0.00	0.78	0.61	0.48	0.38	0.23
	0.02	0.87	0.75	0.65	0.56	0.42
0.05	0.04	0.95	0.91	0.87	0.83	0.75
	0.06	1.05	1.10	1.15	1.21	1.33
	0.08	1.15	1.33	1.53	1.76	2.33
	0.10	1.26	1.59	2.01	2.54	4.04
	0.00	0.62	0.39	0.24	0.15	0.06
	0.02	0.69	0.47	0.32	0.22	0.10
0.10	0.04	0.76	0.57	0.43	0.33	0.19
	0.06	0.83	0.69	0.57	0.48	0.33
	0.08	0.91	0.83	0.76	0.69	0.58
	0.10	1.00	1.00	1.00	1.00	1.00
	0.00	0.50	0.25	0.12	0.06	0.02
	0.02	0.55	0.30	0.17	0.09	0.03
0.15	0.04	0.60	0.37	0.22	0.13	0.05
	0.06	0.67	0.44	0.29	0.20	0.09
	0.08	0.73	0.53	0.39	0.28	0.15
	0.10	0.80	0.64	0.51	0.41	0.26

The actual formula used to calculate these factors, Pr, is :

$$Pr = [\ (1+i) \ / \ (1+d) \ ]^n$$

i = rate of inflation (over and above the general rate)
d = discount rate
n = number of years in the future

Annually-recurring costs or benefits

Table D.2 is used to find the discount factor for an annually-recurring cost or benefit. The annual figure should be multiplied by the appropriate factor to give the total cumulative present worth of those costs over the lifecycle.

Table D.2 Discount factors for annually-recurring costs

Discount rate (d)	Inflation rate (i)	Factor Pa for given number of years				
		5	10	15	20	30
0.00	0.00	5.00	10.00	15.00	20.00	30.00
	0.02	5.31	11.17	17.64	24.78	41.38
	0.04	5.63	12.49	20.82	30.97	58.33
	0.06	5.98	13.97	24.67	38.99	83.80
	0.08	6.34	15.65	29.32	49.42	122.35
	0.10	6.72	17.53	34.95	63.00	180.94
0.05	0.00	4.33	7.72	10.38	12.46	15.37
	0.02	4.59	8.56	11.99	14.96	19.75
	0.04	4.86	9.49	13.91	18.12	25.95
	0.06	5.14	10.54	16.20	22.13	34.86
	0.08	5.45	11.71	18.93	27.24	47.82
	0.10	5.76	13.03	22.21	33.78	66.82
0.10	0.00	3.79	6.14	7.61	8.51	9.43
	0.02	4.01	6.76	8.64	9.93	11.43
	0.04	4.24	7.44	9.86	11.69	14.11
	0.06	4.48	8.20	11.30	13.87	17.78
	0.08	4.73	9.05	12.99	16.59	22.86
	0.10	5.00	10.00	15.00	20.00	30.00
0.15	0.00	3.35	5.02	5.85	6.26	6.57
	0.02	3.54	5.48	6.55	7.13	7.63
	0.04	3.74	6.00	7.36	8.19	8.99
	0.06	3.94	6.56	8.31	9.47	10.76
	0.08	4.16	7.20	9.41	11.03	13.08
	0.10	4.38	7.90	10.71	12.96	16.20

The actual formula used to calculate these discount factors, Pa, for annually recurring costs is :

$$Pa = \frac{x\,(1-x^n)}{(1-x)} \quad \text{where } x = (1+i)\,/\,(1+d)$$

i = rate of inflation (over and above the general rate)
d = discount rate
n = the number of years for which the payment is made.

E.1 Diesel Pumping

Introduction to diesel pumps

This appendix is intended to give the reader enough information to be able to perform an approximate life-cycle cost analysis of diesel pumping. It is only included to give a rough comparison with wind pumping, and should not be used as justification for installing a diesel pump. Further reading on diesel pumping can be found in *Water pumping devices*, (see bibliography).

Diesel technology has the advantage that it is well known and understood all over the world, and so procurement, installation and maintenance are not generally a problem. Diesel pumps will usually be oversized for most small applications, and so even the smallest, rated at about 2.5kW (1.3kW hydraulic power), will probably only need to be run for a short time each day. Short-term water

storage is not a necessity, as water can be pumped at any time on demand, but a tank is still advisable for village water supply applications. The design month is simply that with the highest water demand.

One of the main factors in deciding on the selection of a diesel pump is the cost, quality and availability of fuel. Supplies can often be spasmodic and unreliable in remote areas. Also, note that in an economic appraisal, the international market price of diesel fuel would be used. For a financial assessment the local price should be used, but this may vary widely depending on location.

An attendant is usually needed with a diesel pumping system, and maintenance needs to be frequent.

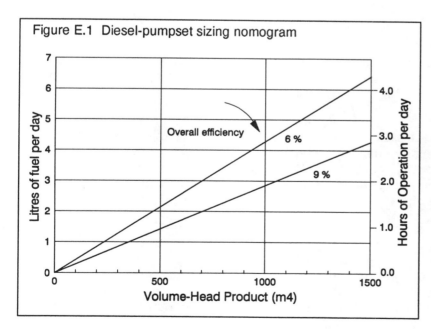

Figure E.1 Diesel-pumpset sizing nomogram

Sizing a diesel pump

As almost all diesel pumps will be oversized for small rural applications, we can select the smallest practical unit. Most pumps will have an overall efficiency between 6 and 9 %, and so the sizing essentially consists of calculating the daily fuel consumption. Using the graph in Figure E.1, it is possible to calculate the number of hours of operation of a diesel pump as a function of volume head product. The second scale on the vertical axis gives a reading directly in litres of fuel used per day. This graph has been constructed for an engine using 1.5 litres of fuel per hour. Thus the number of litres of fuel per day can easily be found if the hydraulic duty in m^4 is known. This can be turned into an annual fuel consumption figure.

Note : Diesel pumps are often rated in Horsepower, HP, where 1 HP is about 0.75 kW.

Costs

Table E.1 gives some figures for guidance on the costs associated with diesel pumping. A pumpset for village water supply is likely to be of a higher quality than for irrigation pumping, and the price will be correspondingly higher. Like operation and maintenance costs, fuel costs are an annual expenditure, and so their contribution to the total life-cycle cost must be calculated using the discount factors as described in Chapter 7.

Table E.1 Diesel pumping costs (in US$)	
Capital Cost	$ 600 to $1000 for a 2.5 kW pumpset
Installation	10 % of diesel capital cost
Pipework	5 $/m
Storage (If necessary)	60 to 150 $/m³
Borehole drilling	60 to 200 $/m
Well-digging	5 to 15 $/m
Engine lifetime	10 years
Pump lifetime	10 years
Fuel (International price - non taxed)	0.30 $ per litre
Fuel (Local price)	0.50 to 2.00 $ per litre
Operation	1 to 2 $ per man day
Maintenance	200 $/year

Note: Many economists forecast that in future years the price of diesel fuel is likely to increase at around 3 % above the general rate of inflation. Therefore you may wish to use i = 0.02 or 0.04 when finding the discount factor from Table D.2 in appendix D.

E.2 Handpumping

Introduction to handpumps

Handpumps are the most widely-used water-lifting devices throughout the developing world. They are renowned for their reliability, simplicity and ease of repair using local technologies.

However, their capacity is limited by the maximum rate of working sustainable by their human operators. Obviously, valuable man hours are also used in pumping water, and the cost of this labour must be attributed to the life-cycle cost of the pump.

It is also worth noting that, in the case of handpumping from boreholes, when one pump is operating at maximum daily capacity, installation of another pump will often require another borehole to be drilled. This

can be extremely costly, and in those circumstances other means of pumping are worth consideration. It may possible to fit more than one pump over a shallow open well, as long as the well can support the increased rate of water extraction.

Except at very low heads and volume requirements, it is unlikely that handpumping will be suitable for irrigation purposes. Note that storage tanks are not normally used with handpumps, as water is pumped on demand.

Sizing the system

Handpumps tend to be of a standard size, which is, of course, related to the normal rate of working of its operator. Thus the system is effectively sized by determining how many person-hours per day are necessary to perform the required hydraulic duty. This can be calculated from the diagram in Figure E.2.

From the water lift (on the horizontal axis) and the daily volume requirement (on the vertical axis) the number of people required can be found. This assumes that a person can produce 60 watts of power, and that the hand-pump efficiency is 60 %. Assuming the working day to be 8 hours this requires an extra pump for every 8 additional person-hours.

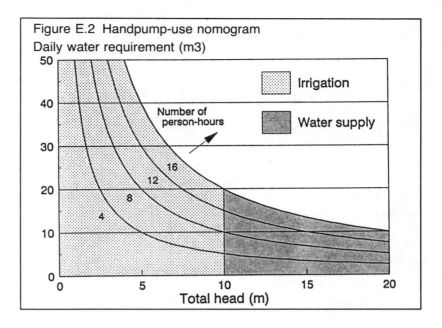

Figure E.2 Handpump-use nomogram

Table E.2	Handpump system costs	(in US$)

Capital cost per pump:
15 m pump $700
25 m pump $1200
40 m pump $1900

Installation .... 12% of hardware capital cost
Lifetime of pump .... 5 years
Borehole Drilling .... 60 to 200$/m
Well-digging .... 5 to 15$/m
Maintenance .... 15 %/year of installed hardware cost
Operation .... 1 to 2$/person-day

Costs

Table E.2 gives some indication of the costs associated with installing and using a handpump. Because many handpumps are manufactured in developing countries, and labour rates may vary, it is important to use local data if available.

Note that the cost of borehole drilling will tend to dominate the capital costs. There will therefore be no economies of scale if another borehole is required for each pump.

The factors involved in a water supply programme are not purely technical. For a project to succeed, its interaction with the social and economic structure of the community as a whole must be considered, and the design and implementation of any programme must reflect this.

User participation

Local involvement is the cornerstone to the development of a successful water supply project. It is important that users feel a sense of responsibility for the system, and local people, be they communities, agencies or private individuals, should be involved in every aspect of water project design, implementation and management.

The following issues must be addressed if projects are to be successful:

- Recipients, whether private individuals, communities or other organizations, must be involved in defining the need for water supply at the start of the project;

- They should be actively involved in establishing the means and rights of access to water supplies;

- They should be involved in site and technology selection;

- They should develop their own systems for maintenance and management of water supply systems once they are installed.

Organization and management

The questions that should be initially addressed are for whom and for what purpose is the water supply system intended. Once these questions are answered then the questions of organization, payment and management must be considered. The scale of the project is also important in this respect, as the design differences are not merely technical, but also centre on organization and responsibility.

A feasibility study to determine the socio-economic impact of a water supply project should be carried out to assess the viability of the project.

Operational costs

Assessing initial capital costs is generally fairly straightforward, but recurrent cost (e.g. maintenance, spares, operation, etc) can present more of a problem. Too often donors will contribute capital costs, but leave the recurrent costs to the recipients without consideration of what they can (or are willing to) afford. Recurrent costs may be unbearable in many circumstances with certain technologies and unworkable unless the recipient is totally committed and able to provide them.

Payment

Community participation in a variety of forms is essential to the success of a project. Many peri-urban and rural people in developing countries pay for water from a variety of sources, and costs and payments vary widely from one group to the next. Virtually all community schemes find it necessary to fix some form of payment, tariff, charge or contribution for access to developed water supplies. It has become a generally accepted rule that payment of some sort is essential, not only to cover operational and recurrent costs, but also to ensure some sense of ownership and responsibility for the water source. Problems inevitably arise when the issue of payment is left to chance.